Haruto

Akari

Sara

Yu

Study with your Friends!

How do we learn mathematics?

Find the ?

Based on the problem you find in your daily life or what you have learned, let's come up with a purpose.

The first problem of the lesson is written. On the left side, what you are going to learn from now on through the problem is written.

Purpose

When you see the problem and think that you "want to think", "want to represent", "want to know", and "want to explore", that will be your "purpose" of your learning. You can find the purpose not only at the beginning of the lesson but in various timings and settings.

You can check your understanding and try more using what you have learned.

①

Let's try this problem first.

The starting point

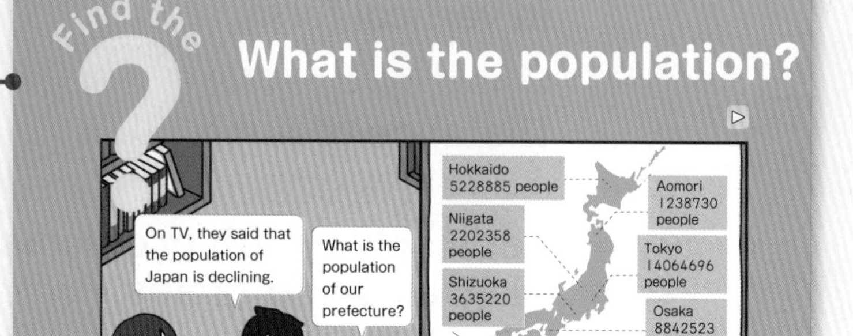

Find the ?
What is the population?
On TV, they said that the population of Japan is declining.
What is the population of our prefecture?
Hokkaido 5228885 people
Aomori 1238730 people
Niigata 2202358 people
Tokyo 14064696 people
Shizuoka 3635220 people
Osaka 8842523 people
Kagoshima

What you have learned tday

Large Numbers

Let's explore how to express numbers **and their** structure.

1 Large Numbers

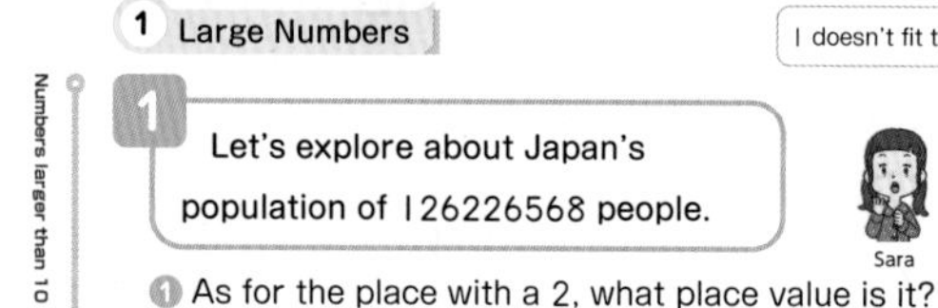

Numbers larger than 10 mil
Let's explore about Japan's population of 126226568 people.

I doesn't fit the table...
Yu
I can read the numbers except the highest place.
Sara

① As for the place with a 2, what place value is it?

Want to know

Purpose How should we read numbers larger than 100 billion?

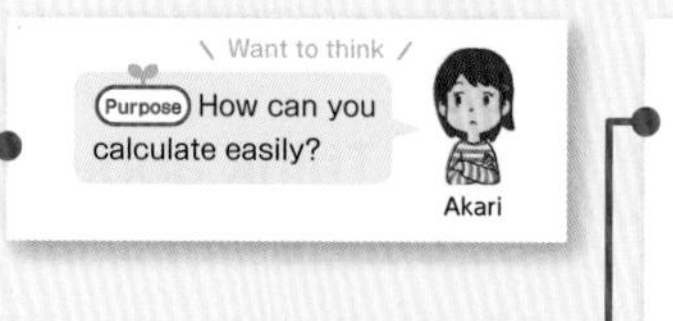

Want to think
Purpose How can you calculate easily?
Akari

2 Let's read the following num

① 9 4600 0000 0000km

(The distance which light travels in o

1 Let's calculate the following in vertical form.

① $54 \div 2$ ② $68 \div 4$ ③ $84 \div 3$ ④ $74 \div 2$

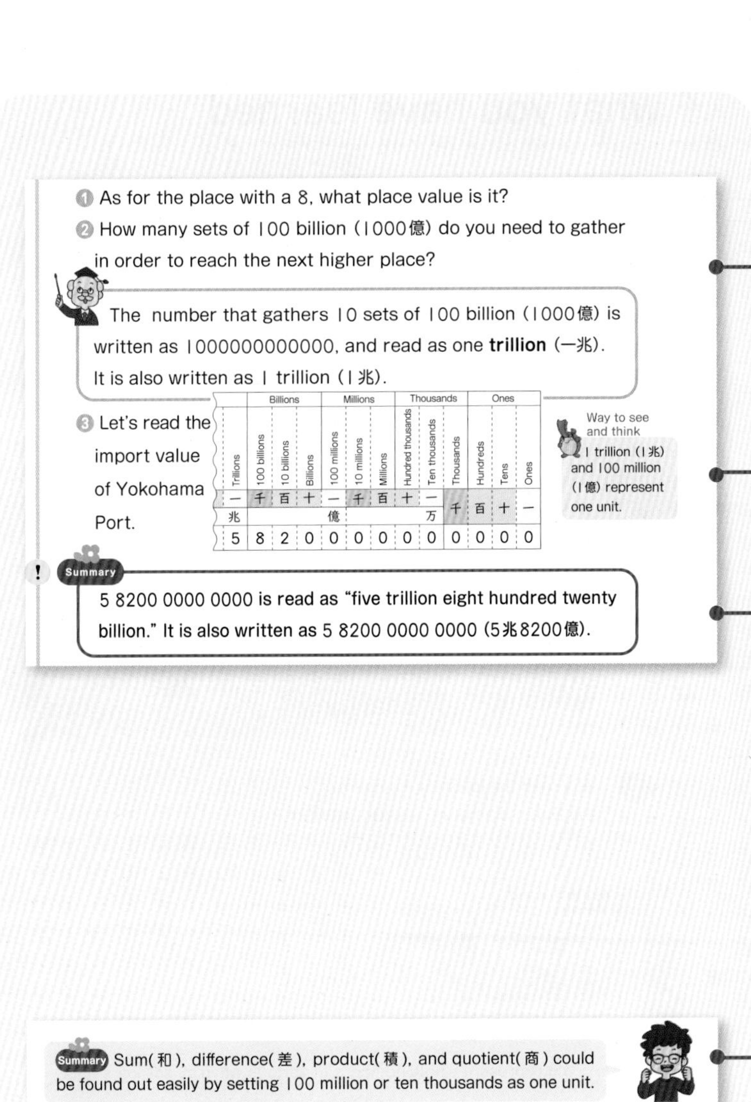

❶ As for the place with a 8, what place value is it?

❷ How many sets of 100 billion (1000億) do you need to gather in order to reach the next higher place?

The number that gathers 10 sets of 100 billion (1000億) is written as 1000000000000, and read as one **trillion** (一兆).
It is also written as 1 trillion (1兆).

❸ Let's read the import value of Yokohama Port.

Trillions	100 billions	10 billions	Billions	100 millions	10 millions	Millions	Hundred thousands	Ten thousands	Thousands	Hundreds	Tens	Ones
一	千	百	十	一	千	百	十	一	千	百	十	一
兆				億				万				
5	8	2	0	0	0	0	0	0	0	0	0	0

(Billions, Millions, Thousands, Ones)

Way to see and think
1 trillion (1兆) and 100 million (1億) represent one unit.

! Summary
5 8200 0000 0000 is read as "five trillion eight hundred twenty billion." It is also written as 5 8200 0000 0000 (5兆8200億).

Summary Sum(和), difference(差), product(積), and quotient(商) could be found out easily by setting 100 million or ten thousands as one unit.

Haruto

You will learn important words and rules from the doctor.

These are the Way to See and Think Monsters which you can find through solving the problem. → See page 8 for more details.

The rules you could find through learning the new content are summarized.

? Can we multiply or divide by 10 for numbers larger than 100 million?

?

You can find problems that will lead you to further learning.

That's it! **Number grouped by 3 digits**

In Japan, every 4 digit place value has its own way of reading for large numbers.

9387 兆 | 4160 億 | 2571 万 | 0364

But in our surroundings, there are many things which are divided by a "," every 3 digits. This is because in Languages like English, reading the digits are assigned for every 3 digits.

9,387,416,025,710,364

That's it!

You can deepen what you have learned.

Summarizing and Reflecting what you have learned

This page is for reflecting on what you can do based on what you have learned.

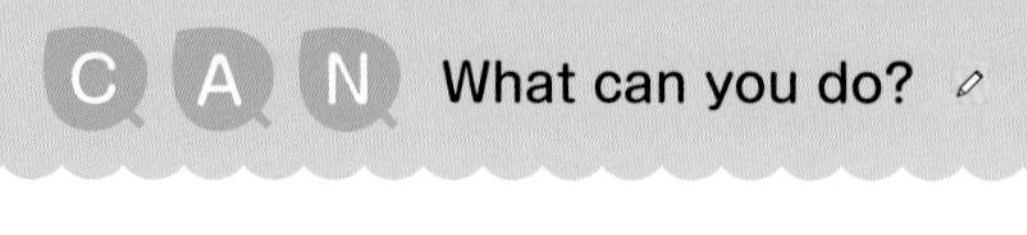

☐ We can read large numbers. → pp.13 ~ 16

1 Let's read the following numbers.
① the distance from the Sun to the Earth 149600000km

You will talk about which "Way to See and Think Monsters" you found in the process of learning.

Which "Way to See and Think Monster" did you find in " 1 Large Numbers"?

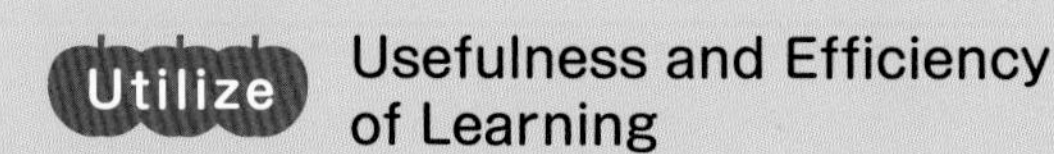

Utilize Usefulness and Efficiency of Learning

This page is for trying to solve a wide variety of problems based on what you have learned.

Utilize Usefulness and Efficiency of Learning

1 Let's read the following numbers.
① the Sun's diameter 1390000000m

With the Way to See and Think Monsters... Let's Reflect!

This page is for reflecting on what you have learned with the "Way to See and Think Monsters."

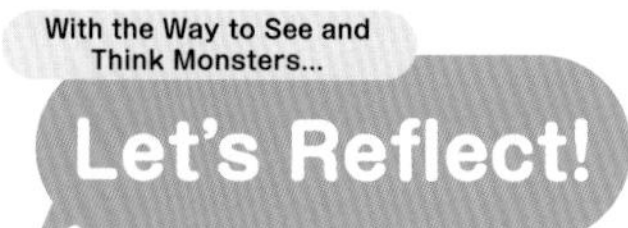

Let's reflect on which monster you used while learning " 1 Large Numbers."

Unit
We could represent whole numbers in various ways by setting a unit.

? Solve the ? Want to Connect

This page is for solving problems based on what you have learned. Moreover, it is for trying to find the next "?" that connects to further learning.

Solve the ? We could use the same way to represent whole numbers when the numbers get large.

Want to Connect How about when the number gets smaller?

This page is for reflecting on what you have learned, and connecting them to your further learning.

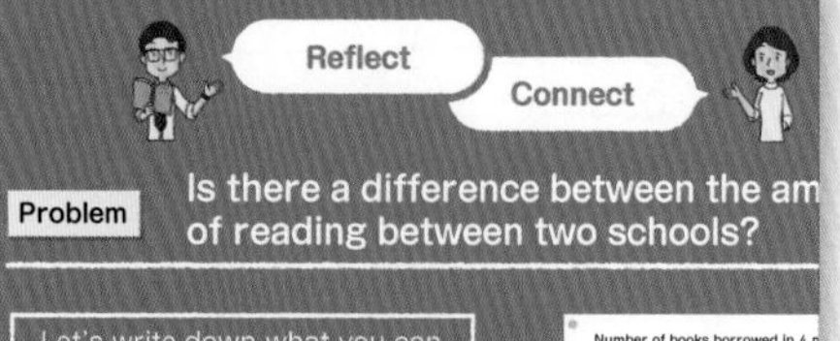

This page is for reviewing areas where you have difficulties to solve the problem, or are likely to make mistakes.

Haruto

 Reflect Connect
Problem Is there a difference between the am of reading between two schools?

Let's write down what you can read from the bar graph and

Number of books borrowed in 4 m
(books) 1000 ■ School A ■ Sc

 Math Patrol

Utilizing Math for SDGs

The Sustainable Development Goals (SDGs) are a set of goals that we aim to achieve in order to create a world where we can live a life of safety and security.
This page will help you think about what you can do for society and the world through math.

About the QR Code

Some of the pages include the QR code which is shown on the right.

▷ … You can learn how to draw a diagram and how to calculate by watching a movie.

… You can learn by actually moving and operating the contents.

… You can learn by reflecting on what you have learned previously in your previous grades.

… You can utilize it to know the solution to the problems that you couldn't find out the answer, or to try various problems.

… You can deepen your learning by actually looking at the materials including the website.

Dear Teachers and Parents

This textbook has been compiled in the hope that children will enjoy learning through acquiring mathematical knowledge and skills.
The unit pages are carefully written to ensure that students can understand the content they are expected to master at that grade level.
In addition, the "More Math!" section at the end of the book is designed to ensure that each student has mastered the content of the main text, and is intended to be handled selectively according to the actual conditions and interests of each child.
We hope that this textbook will help children develop an interest in mathematics and become more motivated to learn.

The sections marked with this symbol deal with content that is not presented in the Courses of Study for that grade level, thus do not have to be studied uniformly by all children.

QR codes are used to connect to Internet content by launching a QR code-reading application on a smartphone or tablet and reading the code with a camera.
The QR Code can be used to access content on the Internet.

https://r6.gakuto-plus.jp/s4a01

Note: This book is an English translation of a Japanese mathematics textbook. The only language used in the contents on the Internet is Japanese.

【Infectious Disease Control】
In this textbook, pictures of activities and illustrations of characters do not show children wearing masks, etc., in order to cultivate children's rich spirit of communicating and learning from each other. Please be careful to avoid infectious diseases when conducting classes.

Becoming a Writing Master

The notebook can be used effectively.

- To organize your own thoughts and ideas
- To summarize what you have learned in class
- To reflect on the what you have learned previously

Let's all try to become notebook masters.

Write today's date. → April 13th

Write the problem of the day that you must solve. → Problem

Each child gets 2 sticks of 3 dumplings. How many dumplings are needed for 4 children?

Write your ideas or what you found about the problem. → ○ My idea

Tried to find out from the number for each child.

The number for each child is 3 dumplings, and there are 2 sticks, so $3 \times 2 = 6$.
The number for each child is 6 dumplings.
There are 4 children, so $6 \times 4 = 24$.
Answer: 24 dumplings

Write the classmate's ideas you consider good. → ○ Yu's idea

Tried to find out from the number of the total number of sticks.

Each child gets 2 sticks, and there are 4 children, so $2 \times 4 = 8$. The total number of sticks is 8.
Each stick has 3 dumplings, so $3 \times 8 = 24$. Answer: 24 dumplings

Tips for Writing ❶

By finding the "Way to See and Think Monsters," it will connect you to what you have learned previously.

Tips for Writing ❷

It's easier to understand important things and new words when you write them in color.

Tips for Writing ❸

By writing down what you would like to try more, it will lead you to further learning.

〈Purpose〉 Do we get the same answer even if we change the order of multiplication?

$(3 \times 2) \times 4 = 24$
$3 \times (2 \times 4) = 24$
↓
$(3 \times 2) \times 4 = 3 \times (2 \times 4)$

Tips for Writing ❶

Did I have a similar case before?

Same way

There was a similar rule when I learned addition.
$(3 + 2) + 4 = 3 + (2 + 4)$

〈Summary〉

In multiplication, the answer will be the same even if the order of multiplication is changed. → Rule of Associativity

Tips for Writing ❷

〈Reflection〉
Summarizing in one math sentence makes it easier to understand what kind of ideas were used in the calculations.

〈What I want to do next〉
I want to explore whether there are other rules of multiplication.

Tips for Writing ❸

Let's write down what you thought while thinking about the solution of the problem as "purpose."

Summarize what you have learned today.

Reflect on your class, and write down the followings;
- What you learned.
- What you found out.
- What you can do now.
- What you don't know yet.

While learning mathematics...

Based on what I have learned previously...

Why does this happen?

There seems to be a rule.

You may be in situations like above. In such case, let's try to find the "Way to See and Think Monsters" on page 9. The monsters found there will help you solve the mathematics problems. By learning together with your friend and by finding more "Way to See and Think Monsters," you can enjoy learning and deepening mathematics.

What can we do at these situations?

I think I can use 2 different monsters at the same time...

→ You may find 2 or 3 monsters at the same time.

I came up with the way of thinking which I can't find on page 9.

→ There may be other monsters than the monsters on page 9.
Let's find some new monsters by yourselves.

Now let's open to page 9 and reflect on the monsters you found in the 3rd grade. They surely will help your mathematics learning in the 4th grade!

Representing ways of thinking in mathematics

Way to See and Think Monsters

Unit

If you set the unit...

Once you have decided one unit, you can represent how many using the unit.

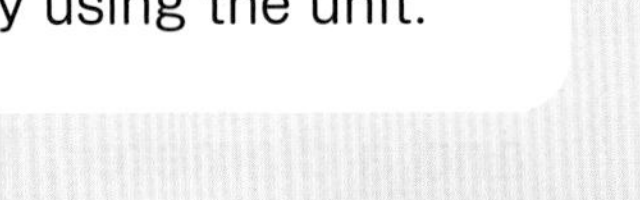

Summarize

If you try to summarize...

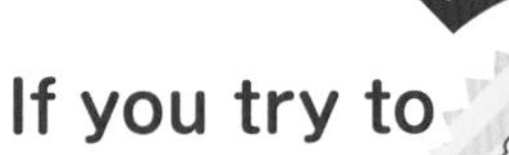

It makes it easier to understand if you summarize the numbers or summarize in a table or a graph.

Other Way

If you represent in other ways...

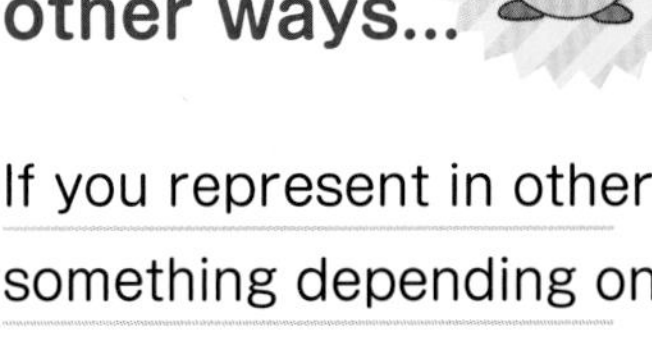

If you represent in other something depending on your purpose, it is easier to understand.

Align

If you try to align...

You can compare if you align the number place and align the unit.

Change

If you try to change the number or the figure...

If you try to change the problem a little, you can understand the problem better or find a new problem.

Divide

If you try to divide...

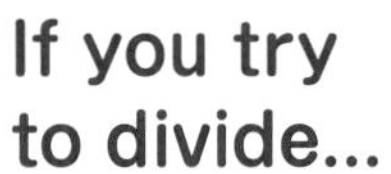

Decomposing numbers by place value and dividing figures makes it easier to think about problems.

Why

You wonder why?

Why does this happen? If you communicate the reasons in order, it will be easier to understand for others.

Rule

Is there a rule?

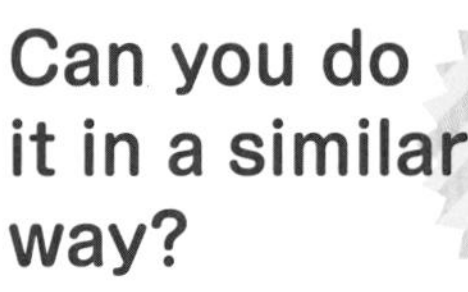

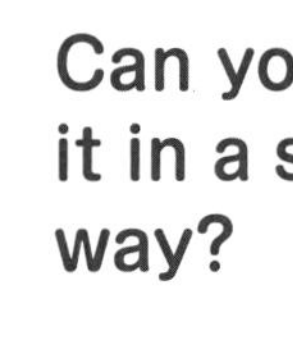

By examining, you can find rules and think using rules.

Same Way

Can you do it in a similar way?

If you find something the same or similar to what you have learned, you can understand.

Ways to think learned in the 3rd grade

Numbers and Calculations

Align **Same Way**

In addition and subtraction of large numbers, we can do the same by aligning the place value.

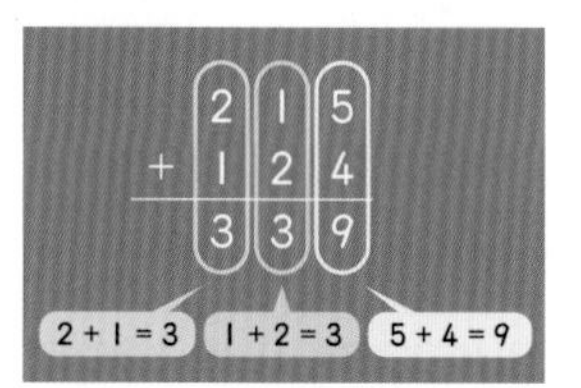

Divide

Multiplicand is calculated divided by each place value.

$$\begin{array}{r} 23 \\ \times\ 12 \\ \hline 46 \\ 23 \end{array}$$

Same Way

Division can be used in both cases, to find out the number for each and the number of unit.

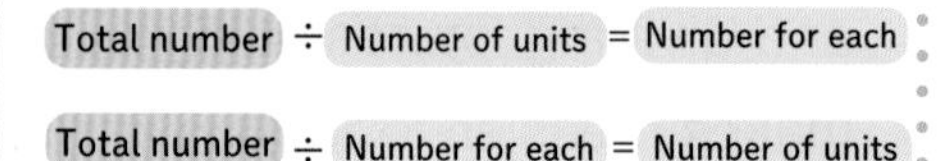

Total number ÷ Number of units = Number for each
Total number ÷ Number for each = Number of units

Change

Change the dividend and explore the relationship between the answer and the remainder.

Dividend		Divisor		Answer	Remainder
12	÷	4	=	3	
11	÷	4	=	2	remainder 3
10	÷	4	=	2	remainder 2
9	÷	4	=	2	remainder 1
8	÷	4	=	2	

Unit

The size of the unit is represented in fractions using equally divided parts.

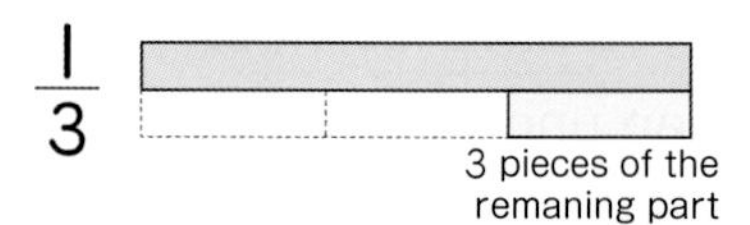

Unit

Decimals, like whole numbers, can be represented in numbers by considering each decimal place as one unit.

Tens	Oens	Tenths
2	5 .	7

Shapes

Why

We can identify a circle by understanding the meaning of the circle.

radius
center
radius
radius

Same Way

Circles and spheres both have a center, radius and a diameter. The length of the diameter is always twice the length of the radius.

diameter
radius
center

Summarize

Classify triangles by the length of the sides.

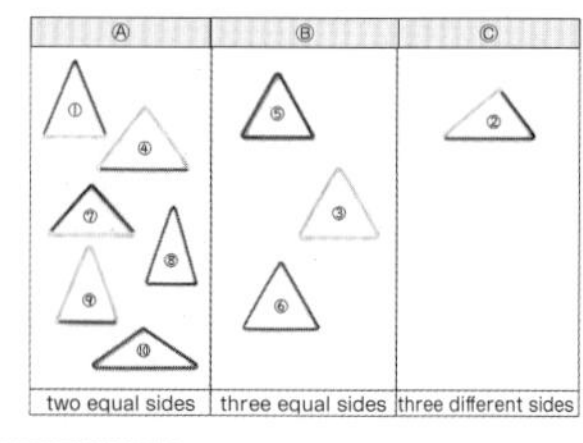

two equal sides
three equal sides
three different sides

Why

Can explain why we can draw triangles.

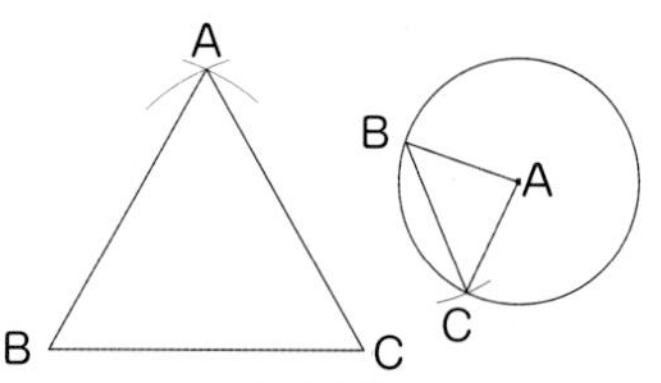

Data

Measurements

Other Way

Represent time or hours in a number line.

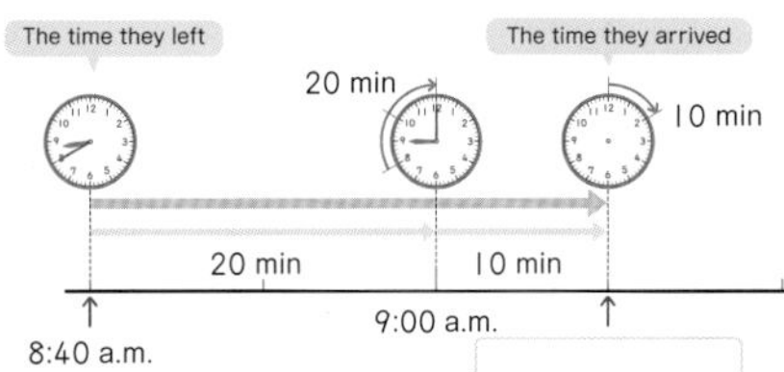

Unit

Representing weight and length by setting 1g and 1kg (weight), and 1cm, 1m, and 1km (length) as one unit.

Same Way

Length, amount, and weight have a similar relationship in the rule of the units.

				kilo k				deci d	centi c	milli m
Length				1km			1m		1cm	1mm
Amunt of watet				1kL			1L	1dL		1mL
Weight	1t			1kg			1g			1mg

1000 times 1000 times 100 times 10 times 1000 times

Other Way

Represent in a bar graph or a table to make it easier to understand.

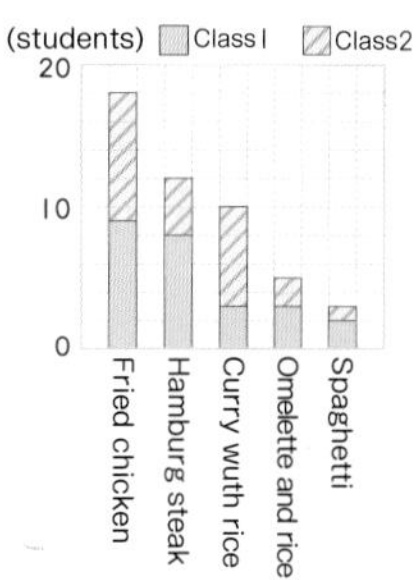

Unit

Can make graphs easier to understand by changing the amount of one scale of the graph.

(cm) ① 60 40 20 0

② (L) 30 20 10 0

Summarize

Summarize in one table to compare easily.

Menu \ Class	Class1	Class2	Total
Fried chicken	9	9	18
Hamburg steak	8	4	12
Curry and rice	3	7	10
Omelette with rice	3	2	5
Spaghetti	2	1	3
Total	25	23	48

Find the ?

What is the population?

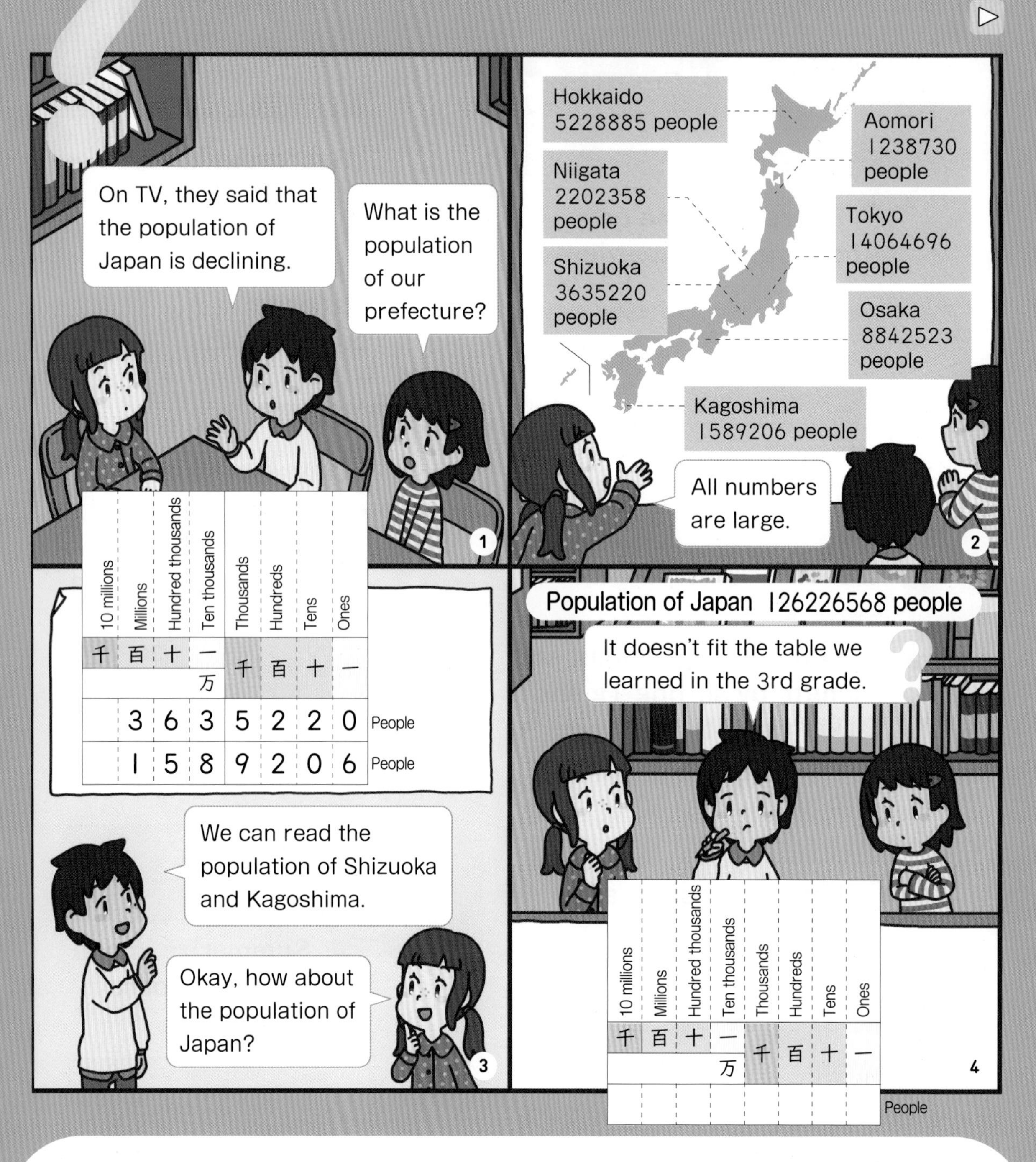

On TV, they said that the population of Japan is declining.
What is the population of our prefecture?
1
Hokkaido 5228885 people
Aomori 1238730 people
Niigata 2202358 people
Tokyo 14064696 people
Shizuoka 3635220 people
Osaka 8842523 people
Kagoshima 1589206 people
All numbers are large.
2
10 millions | Millions | Hundred thousands | Ten thousands | Thousands | Hundreds | Tens | Ones
千 | 百 | 十 | 一 万 | 千 | 百 | 十 | 一
3 6 3 5 2 2 0 People
1 5 8 9 2 0 6 People
We can read the population of Shizuoka and Kagoshima.
Okay, how about the population of Japan?
3
Population of Japan 126226568 people
It doesn't fit the table we learned in the 3rd grade.
10 millions | Millions | Hundred thousands | Ten thousands | Thousands | Hundreds | Tens | Ones
千 | 百 | 十 | 一 万 | 千 | 百 | 十 | 一
People
4

Want to know

Purpose How can we read the population of Japan?

1 Large Numbers

Let's explore how to express numbers and their structure.

1 Large Numbers

Numbers larger than 10 million ↓

I (one) doesn't fit the table...

I can read the numbers except the highest place.

1 Let's explore about Japan's population of 126226568 people.

① As for the place with a 2, what place value is it?

② How many sets of 10 million (千万) do you need to gather in order to reach the next higher place?

The number that gathers 10 sets of 10 million (千万) is written as 100000000, and read as one **hundred million** (一億). It can be also written as 100 million (1億).The place value that follows the ten millions place is the **hundred millions place**.

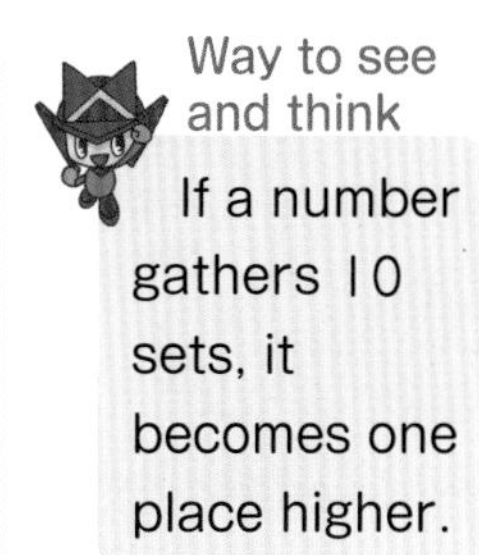

Way to see and think
If a number gathers 10 sets, it becomes one place higher.

③ Let's read the population of Japan.

Millions			Thousands			Ones			
100 millions	10 millions	Millions	Hundred thousands	Ten thousands	Thousands	Hundreds	Tens	Ones	
一	千	百	十	一	千	百	十	一	
億				万					
1	2	6	2	2	6	5	6	8	People

If the table is written, it becomes easier to read.

Haruto

Way to see and think

100 million (1億) and 10 thousand (1万) represent one unit.

Summary

126226568 is read as one hundred twenty-six million two hundred twenty six thousand five hundred sixty-eight. (1億2622万6568).

1 Let's write the following numbers in numerals.

① The number that gathers 10 sets of 100 million (1億) is ______.

② The number that gathers 10 sets of 1 billion (10億) is ______.

③ The number that gathers 100 sets of 100 million (1億) is ______.

Way to see and think

Gathering 10 sets of 10 thousand (1万) makes 100000 (10万).

I wonder if the hundred millions place works the same as the ten thousands place.

Akari

2 Let's write and read the population of Brazil, India, U.S.A., and the world.

It's easy to start writing from the ones place in the table.

I also want to try to read the world's population.

Yu

	Billions			Millions			Thousands			Ones			
	100 billions	10 billions	Billions	100 millions	10 millions	Millions	Hundred thousands	Ten thousands	Thousands	Hundreds	Tens	Ones	
	千	百	十	一 億	千	百	十	一 万	千	百	十	一	
Brazil				2	1	2	5	5	9	0	0	0	People
India													People
U.S.A													People
World													People

7795000000 people
World's population

46755000 people
Spain

1380004000 people
India

331003000people
U.S.A.

126227000 people
Japan

212559000 people
Brazil

53771000 people
Kenya

25500000 people
Australia

3 Let's look for a number that uses the hundred millions place from your surroundings.

? Are there numbers larger than 100 billion?

2

The following amount of money is the import value of Yokohama Port in 2020.
5 8200 0000 0000 yen
Let's explore this number.

(Yokohama City, Kanagawa Prefecture)

Sara: There is a place value higher than the 100 billions place.

Haruto: Is there another place value?

\ Want to know /

? **Purpose** How should we read numbers larger than 100 billion?

① As for the place with a 8, what place value is it?

② How many sets of 100 billion (1000億) do you need to gather in order to reach the next higher place?

The number that gathers 10 sets of 100 billion (1000億) is written as 1000000000000, and read as one **trillion** (一兆).
It is also written as 1 trillion (1兆).

③ Let's read the import value of Yokohama Port.

	Billions			Millions			Thousands			Ones		
Trillions	100 billions	10 billions	Billions	100 millions	10 millions	Millions	Hundred thousands	Ten thousands	Thousands	Hundreds	Tens	Ones
一	千	百	十	一	千	百	十	一	千	百	十	一
兆				億				万				
5	8	2	0	0	0	0	0	0	0	0	0	0

Way to see and think
1 trillion (1兆) and 100 million (1億) represent one unit.

! **Summary**

5 8200 0000 0000 is read as "five trillion eight hundred twenty billion." It is also written as 5 8200 0000 0000 (5兆8200億).

④ Let's try to expand the table on ③ until it reaches the quadrillions* place.

* 1 quadrillion is that gathers 1000 sets of 1 trillion.

1 When Akari reads the number 9754800000000, she draws a line every 4 digits like shown below. Let's explain Akari's idea.

9 | 7548 | 0000 | 0000

Akari

Way to see and think

Inside the representing place value of thousands (万), 100 millions (億), and trillions (兆), the ones (一), tens (十), hundreds (百), and thousand (千) becomes repeated.

2 Let's read the following numbers.

① 9 4600 0000 0000km
(The distance which light travels in one year)

② 4078 0000 0000 0000km
(The distance between the Earth and the North Star)

The distance which light travels in a year is called 1 light year.

? Can we multiply or divide by 10 for numbers larger than 100 million?

That's it! Number grouped by 3 digits

In Japan, every 4 digit place value has its own way of reading for large numbers.

9387	4160	2571	0364
兆	億	万	

But in our surroundings, there are many things which are divided by a "," every 3 digits. This is because in Languages like English, reading the digits are assigned for every 3 digits.

9,387,416,025,710,364

3

Let's write and read the numbers that are 10, 100, and 1000 times of 325 6900. Let's write and read the number which is $\frac{1}{10}$ of 325 6900.

? **Purpose** \ Want to represent /

In a whole number, when getting 10 times or getting $\frac{1}{10}$, how does the place value change?

Billions			Millions			Thousands			Ones		
100 billions	10 billions	Billions	100 millions	10 millions	Millions	Hundred thousands	Ten thousands	Thousands	Hundreds	Tens	Ones
千	百	十	一 億	千	百	十	一 万	千	百	十	一
					3	2	5	6	9	0	0

$\frac{1}{10}$ / 10 times / 10 times / 10times / 100 times / 1000 times

1 Let's write the numbers that are 10, 100, 1000 times, and $\frac{1}{10}$ of the following numbers.

Way to see and think: Can we apply the same thing for other numbers as in 3?

① 6 billion (60億) ② 400 thousand (40万) ③ 80 billion (800億)

2 Let's write the following numbers in numerals.

10 times 100 million (1億) is ______.

100 times 100 million (1億) is ______.

1000 times 100 million (1億) is ______.

10000 times 100 million (1億) is ______.

! **Summary**

In any whole number, when getting 10 times, the place value increases one position. When getting $\frac{1}{10}$, the place value decreases one position.

3 Let's fill in each ☐ with a number in the following number lines.

① 0 ☐ 50 million ☐ 100 million ☐

② 0 1 billion ☐ ☐ 10 billion ☐

③ 0 ☐ ☐ 700 billion 1 trillion ☐

Way to see and think
How much does one scale represent?

4 Let's fill in each ☐ with inequality signs.

① 213610000 ☐ 21363000

② 110950000 ☐ 111095000

? Can we represent numbers by only using numbers from 0 to 9 however large the numbers get?

That's it! Numbers with place value higher than 1 quadrillion (1000兆) ADVANCED

1 quadrillion (1000兆) is a very large number. For example, if you count one number each second, it will take you about 30 million (3000万) years to count from 1 to 1 quadrillion (1000兆).

Additionally, there are place values higher than 1 quadrillions (1000兆) place. If you write the number character for "Mu-ryo-tai-su", 68 zeros are aligned after 1.

1	0000	0000	0000	0000	0000	0000	0000	0000	0000	0000	0000	0000	0000	0000	0000	0000	0	0	0	0
mu-ryo-tai-su	fukashigi	nayuta (novemdecillion)	a-sou-gi	gou-ga-sya: in buddhism, from Gou-ga-sya the number system changes to 8-digit system instead of 4-digit system)	1 quindecillion	100 tredecillion (no quattuordecillion)	10 duodecillion	100 decillion	100 nonillion	10 octillion	1 septillion	100 quintillion (no sextillion)	10 quadrillion	1 trillion	100 million (no billion)	10 thousand (no 100 thousand)	Thousand	Hundred	Tens	Ones

1 scale of the number line below is 1mm. Let's examine how long 100000000 would be.

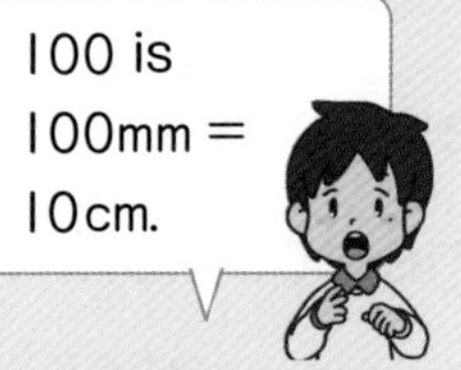

1 trillion is 1000000km. We can go around the earth for around 25 times.

0 10 50 100

2 Structure of whole numbers

1

Using the ten cards from 0 to 9, let's create 10-digit whole numbers.

	Millions			Thousands			Ones		
Billions	100 millions	10 millions	Millions	Hundred thousands	Ten thousands	Thousands	Hundreds	Tens	Ones
十	一	千	百	十	一	千	百	十	一
	億				万				

Akari: On each place value, one number character can be inserted.

Haruto: We can't use 0 in the billions (十億) place.

We can make a lot of whole numbers.

1. Let's make the largest number.
2. Let's make the smallest number.

Want to know

Purpose Can we represent any number using the cards from 0 to 9?

Yu

Any whole number can be written and represented by using the following 10 numerals: 0, 1, 2, 3, 4, 5, 6, 7, 8, 9.

1

Considering 30 9800 0000 0000, let's fill in the ☐.

① It is the sum of 30 sets of 1 trillion (1兆) and ☐ sets of 100 million (1億).

② It is the sum of ☐ sets of 10 trillion (10兆), ☐ sets of 100 billion (1000億), and 8 sets of 10 billion (100億).

③ It's a number that gathers ☐ sets of 100 million (1億).

Way to see and think

As in 1 trillion (1兆) and 100 million (1億), the representation of whole numbers change depending on what is taken as a unit.

	Trillions			Billions				Millions				Thousands		Ones	
Quadrillions	100 trillions	10 trillions	Trillions	100 billions	10 billions	Billions	100 millions	10 millions	Millions	Hundred thousands	Ten thousands	Thousands	Hundreds	Tens	Ones
千	百	十	一	千	百	十	一	千	百	十	一	千	百	十	一
			兆				億				万				
		3	0	9	8	0	0	0	0	0	0	0	0	0	0
10 millions	Millions	Hundred thousands	Ten thousands	Thousands	Hundreds	Tens	Ones								

? Can we add, subtract, multiply, or divide using large numbers?

3 Calculation of large numbers

1 The table on the right shows the number of copies of magazines that have been published in Japan this year. Let's think about these numbers.

Number of published magazines in Japan

Type	Number of copies
Weekly magazine	700000000
Monthly magazine	1800000000

❶ Find out the sum of the number of copies of magazines published weekly and monthly.

ⓐ 700000000 + 1800000000

ⓑ 700 million + 1 billion 800 million

Way to see and think
As for 700000000, if 100000000 is one unit, then it has 7 units. It can be represented as700 million (7億).

Want to think
Purpose How can you calculate easily?

Akari

Yu

There are 10 digits in ⓐ. I might make a mistake the place value.

In the case of ⓑ, I can calculate mentally.

Sara

❷ What is the difference between the number of monthly and weekly publication magazines?

The result of adding numbers is called **sum**. The result of subtracting numbers is called **difference**.

1 Let's find out the following sums and differences.

① Sum of 1 billion 700 million and 2 billion 900 million

② 2 million 350 thousand + 5 million 150 thousand

③ Difference of 23 trillion and 8 trillion

④ 80 billion 700 million − 69 billion 200 million

Words

Sum 【和】 To combine together.

Difference 【差】 Take away.

How much yen is the sum?
How much yen is the difference?
500 million (5億) yen
800 million (8億) yen

2 Let's think about the following problems.

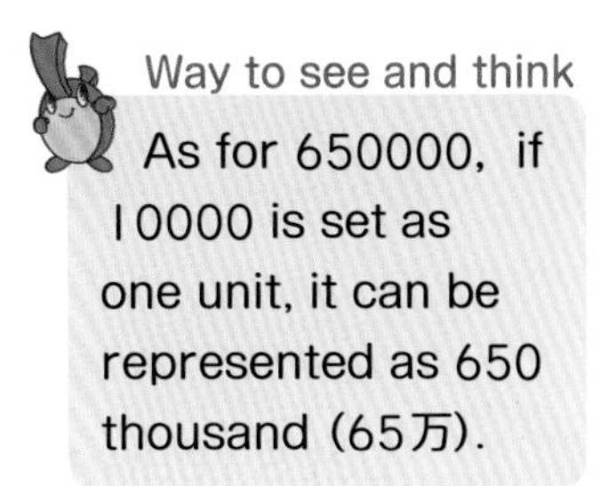

Way to see and think
As for 650000, if 10000 is set as one unit, it can be represented as 650 thousand (65万).

① Yuma's city library has a monthly-budget of 650000 yen to purchase books. How much is the annual budget?

Math Sentence:

② Rin's school budget to buy lunch for 5 days is 350000 yen. How much is the daily fee?

Math Sentence:

The result of multiplying numbers is called **product**. The result of dividing numbers is called **quotient**.

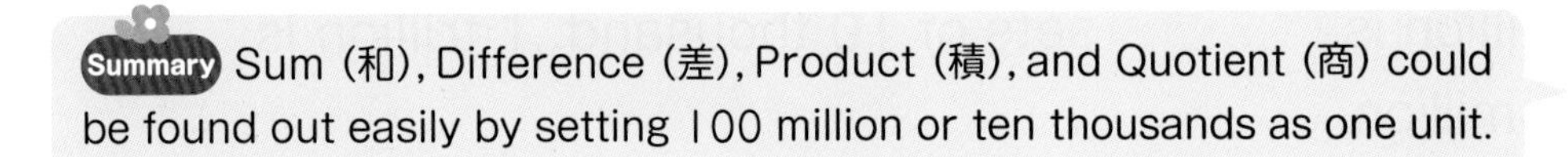

Summary
Sum (和), Difference (差), Product (積), and Quotient (商) could be found out easily by setting 100 million or ten thousands as one unit.

3 Let's find out the following products or quotients.

① 760 thousand × 2

② 26 billion 400 million × 10

③ 8 million 500 thousand ÷ 10

④ 90 trillion ÷ 9

Pile up the same sets.

Measuring for sale.

How many in total?

How many buckets of 10 L can I pour in?

CAN What can you do?

☐ We can read large numbers. → pp.13 ~ 16

1 Let's read the following numbers.

① the distance from the Sun to the Earth 149600000km

② total Budget of the Japanese Government (in 2021) 106609700000000 yen

☐ We can compare the size of the numbers. → p.18

2 Let's fill in each ☐ with inequality signs.

① 9860124700 ☐ 98600124700

② 7049082513 ☐ 7049802513

☐ We understand how large numbers work. → p.19

3 Let's fill in each ☐ with a number or a word.

① The number that gathers 10 sets of 10 million is ☐. The number that gathers 10 sets of 100 billion is ☐.

② 100 million is ☐ sets of 10 thousand. 1 trillion is ☐ sets of 100 million.

③ The digit 7 in the number 72000000000000 represents 7 sets of ☐.

④ The number that is 10 times 180 billion is ☐.

⑤ The number that is $\frac{1}{10}$ of 23 trillion is ☐.

⑥ The number that gathers 20 sets of 10 trillion and 45 sets of 10 billion is ☐.

Supplementary Problems → p.142

Which "Way to See and Think Monsters" did you find in "1 Large Numbers"?

I found "Unit" when I was trying to represent large numbers.
Sara
I found other monsters too!
Akari

Utilize Usefulness and Efficiency of Learning

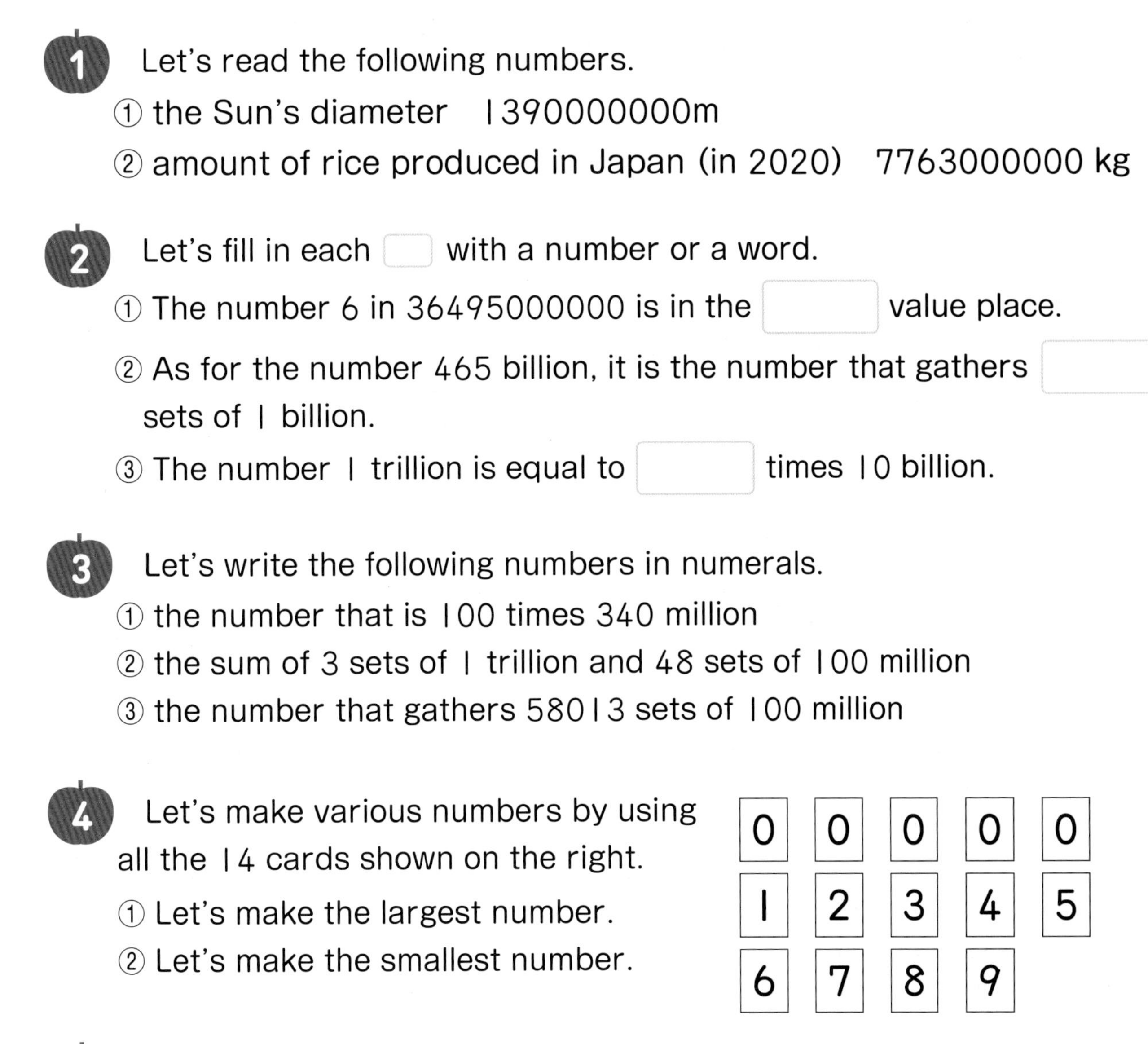

1 Let's read the following numbers.

① the Sun's diameter　1390000000m

② amount of rice produced in Japan (in 2020)　7763000000 kg

2 Let's fill in each ☐ with a number or a word.

① The number 6 in 36495000000 is in the ☐ value place.

② As for the number 465 billion, it is the number that gathers ☐ sets of 1 billion.

③ The number 1 trillion is equal to ☐ times 10 billion.

3 Let's write the following numbers in numerals.

① the number that is 100 times 340 million

② the sum of 3 sets of 1 trillion and 48 sets of 100 million

③ the number that gathers 58013 sets of 100 million

4 Let's make various numbers by using all the 14 cards shown on the right.

① Let's make the largest number.

② Let's make the smallest number.

0	0	0	0	0
1	2	3	4	5
6	7	8	9	

5 Let's calculate the followings.

① 2400000000 + 4500000000

② 38000000000000 − 16000000000000

③ 750000 × 12　　④ 640000 ÷ 8

⑤ 38 billion 300 million + 42 billion 900 million

⑥ 73 million 510 thousand − 3 million 960 thousand

⑦ 5 million 260 thousand × 5　　⑧ 7 billion 200 million ÷ 8

With the Way to See and Think Monsters...

Let's Reflect!

Let's reflect on which monster you used while learning "1 Large Numbers."

Unit

We could represent whole numbers in various ways by setting a unit.

① What tips did we learn to represent the whole numbers?

	Billions			Millions			Thousands			Ones		
Trillions	100 billions	10 billions	Billions	100 millions	10 millions	Millions	Hundred thousands	Ten thousands	Thousands	Hundreds	Tens	Ones
一	千	百	十	一	千	百	十	一	千	百	十	一
兆				億				万				
			1	8	0	0	0	0	0	0	0	0

By setting 1 trillion or 100 million as one unit, we could represent large numbers easily.

Akari

There are eighteen 100 millions in 1800000000.

Yu

Same Way

However large the number is, numbers have the same structure.

② What characteristics do the whole numbers have?

In any whole number, the place value increases to a higher place when getting 10 times of the number.

Sara

	Trillions			Billions				Millions				Thousands		Ones		
Quadrillions	100 trillions	10 trillions	Trillions	100 billions	10 billions	Billions	100 millions	10 millions	Millions	Hundred thousands	Ten thousands	Thousands	Hundreds	Tens	Ones
千	百	十	一	千	百	十	一	千	百	十	一	千	百	十	一
			兆				億				万				

It is the repetition of ones, tens, hundreds, and thousands in the ten thousands, hundred thousands, and millions place.

Haruto

Let's deepen. → p.150

Solve the ?

We could use the same way to represent whole numbers when the numbers get large.

Yu

Want to Connect

How about when the number gets smaller?

Haruto

Find the

Comparing Temperatures?

These are the photos of Niigata City and Naha City in January.
There is snow in Niigata.
It's still January, but cherry blossoms are blooming in Naha.
1
Are the temperatures different in these two cities?
I want to compare the temperature of the two cities.
I want to compare it not only for January but also for the whole year.
2
Can we represent them using a bar graph we learned in 3rd grade?
Are there easier ways to represent them?
3

Purpose

\ Want to know /

What is the good way to represent the comparison of temperatures?

Line Graphs

Let's explore graphs to represent changes.

1 Line graph

1

The following table show the average temperature per month in Niigata City and Naha City. Let's answer the following questions.

The unit to represent temperature is ℃ .

Average temperature per month in Niigata City and Naha City

Month	1	2	3	4	5	6	7	8	9	10	11	12
Niigata City (℃)	5	5	8	10	17	22	24	28	24	16	11	5
Naha City (℃)	19	19	20	20	25	28	29	29	28	26	23	19

❶ Let's look at the table above and compare how the temperature changes and the difference of the temperature per month in two cities.

❷ The bar graph below shows the average temperature per month in Niigata City. Discuss what you notice about the bar graph with your friends.

Average temperature per month in Niigata City

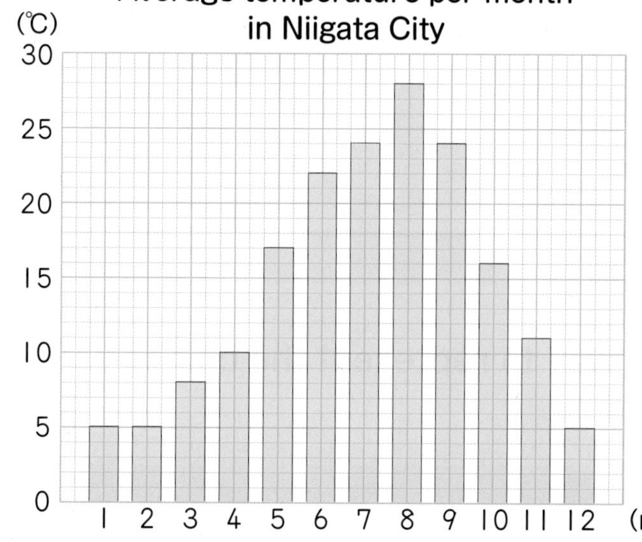

Yu
We can see which month has the highest and the lowest temperature.

Akari
It is a little difficult to understand how the temperature changes. Are there other ways that makes easier to see it?

3 Where should we focus on in order to find out how the temperature changes?

Seeing the top part of the bar graph might be good...
Haruto

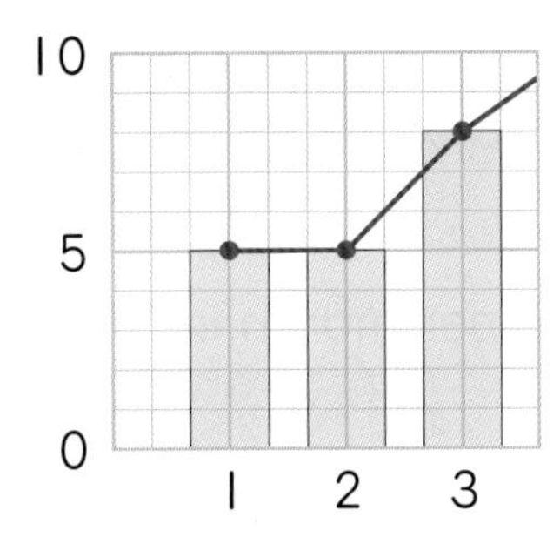

By drawing a point at the end of the bar and connecting them with a line as shown on the left, it makes it easier to find out the changes in temperature.

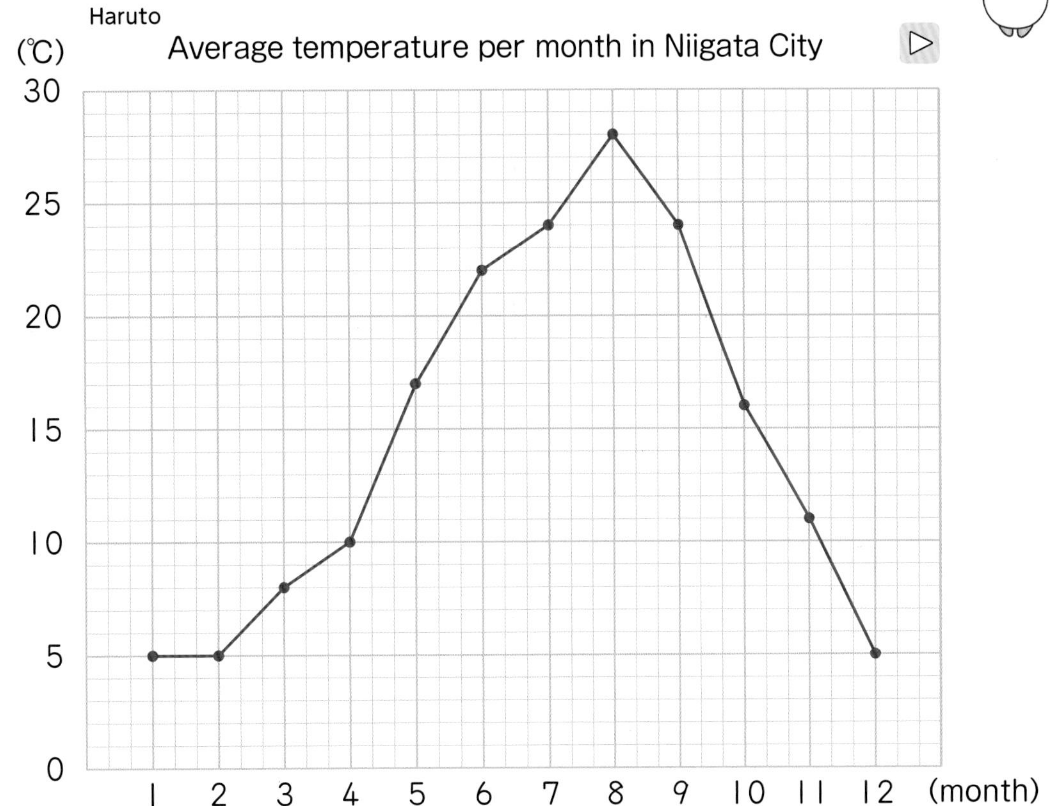

4 What is the temperature (℃) in March? In which month is the temperature 16℃ ?

A graph that represents the rate of change in temperature etc., is called a **line graph**.

? What are the merits of using a line graph?

2

The following line graphs show the average temperature per month in Niigata City and Naha City. Let's compare the temperatures in the two cities.

Average temperature per month in Niigata City and Naha City

Month	1	2	3	4	5	6	7	8	9	10	11	12
Niigata City (℃)	5	5	8	10	17	22	24	28	24	16	11	5
Naha City (℃)	19	19	20	20	25	28	29	29	28	26	23	19

Average temperature per month in Niigata City and Naha City

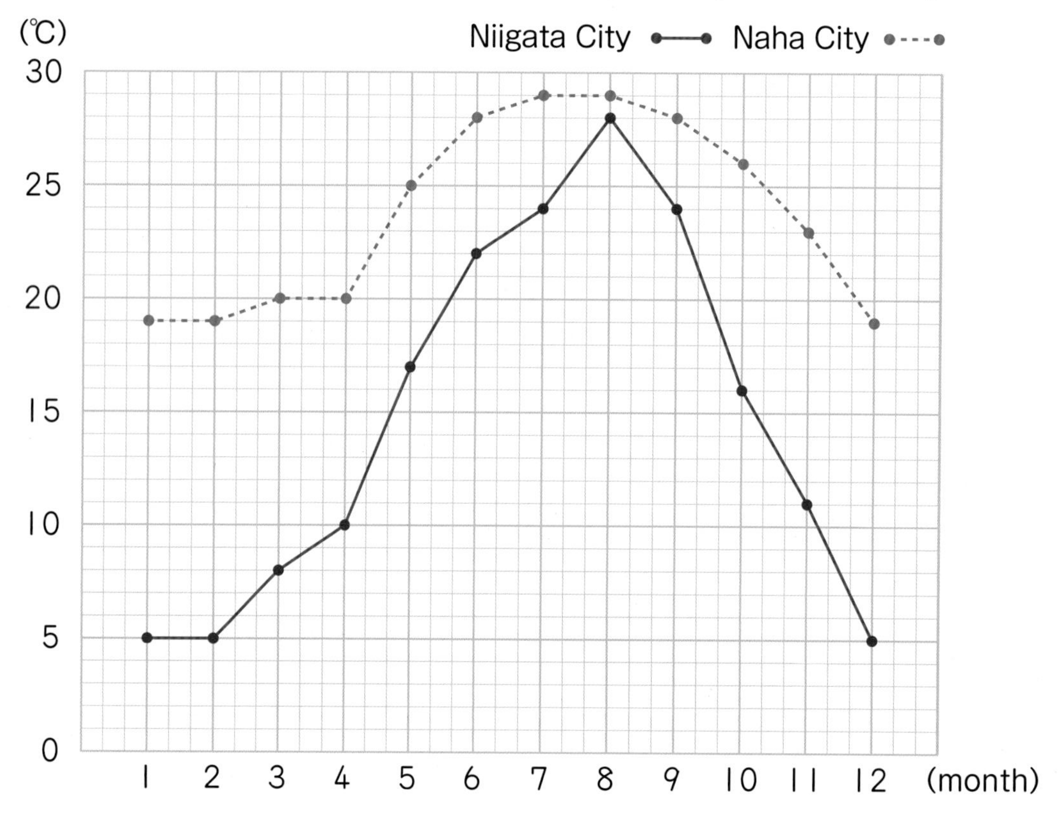

Want to know

? **Purpose** If we put the two line graphs together, what can we find out easily?

① Which month has the highest temperature in each city? How many degrees are they?

② How does the temperature change? Let's compare how the temperature changes in Niigata City and Naha City.

③ Which city has the highest change in temperature between two consecutive months? Name the respective months.

We can see the degree of change depending on the slope of the line.

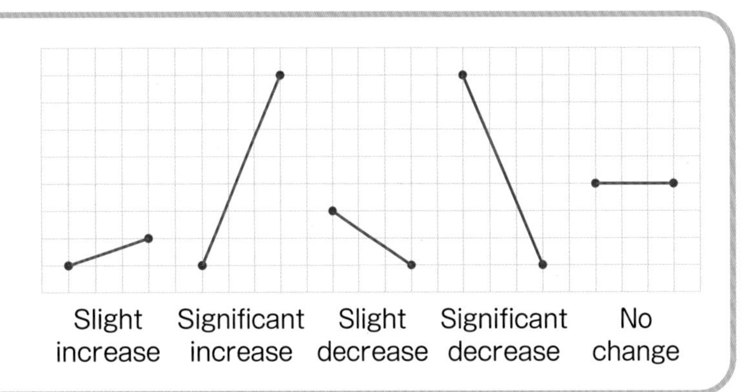

④ Let's discuss the merits of line graphs.

Way to see and think

By using line graphs, we can represent changes of temperature per month easily. By putting the two line graphs together, we can see the difference clearly.

1 Considering situations Ⓐ~Ⓕ, which is appropriate to be represented by a line graph?

Ⓐ your body temperature taken at the same time every day

Ⓑ the type and number of vehicles that passed in front of your school in a period of 10 minutes

Ⓒ the type of favorite fruit of your classmates and the number of the classmates for each type

Ⓓ the temperature recorded every hour at the same place

Ⓔ the height of your classmates

Ⓕ your height measured on your birthday every year

Can we draw a line graph by ourselves?

2 How to draw a line graph

Rurikoji Temple Five-Story Pagoda
(Yamaguchi City, Yamaguchi Pref.)

1 The table below shows how temperature changed in Yamaguchi City. Let's represent this information in a line graph.

Temperature change in Yamaguchi City (April 19th, 2021)

Time (hr)	a.m. 9	10	11	12	p.m. 1	2	3
Temperature (℃)	12	16	18	19	21	22	21

\ Want to try /

Purpose Let's continue drawing the graph based on "How to draw a line graph" below.

Sara

How to draw a line graph 

① Write units for the horizontal and vertical axis.
② Write time digits in the horizontal axis with the same spacing, and write a scale that represents 22 ℃ as the highest temperature in the vertical axis.
③ Based on the table, put a dot for each corresponding time and temperature.
④ Connect the dots with a line.
⑤ Write the title.

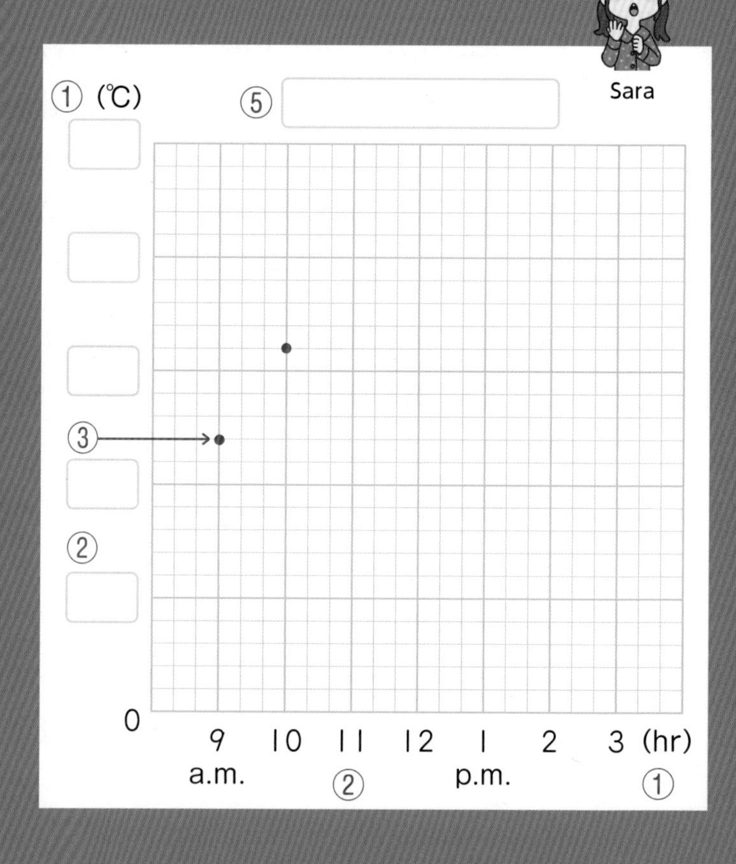

1 Ayumi investigated about how the length of a shadow changed. The following tables show the records of the length of the 10 cm stick's shadow in June and December. Let's represent the tables in a line graph.

Length of the shadow in June

Time (hr)	8	9	10	11	12	13	14	15
Length (cm)	12.1	7.9	4.9	2.8	2.1	3.5	6	9.3

Length of the shadow in December

Time (hr)	8	9	10	11	12	13	14	15
Length (cm)	51	27.8	20	16.8	16.3	18.1	23.1	36.1

① For each month, which is the biggest change in length between two consecutive hours? Name the hours.

② Look at the graph and write down what you found out.

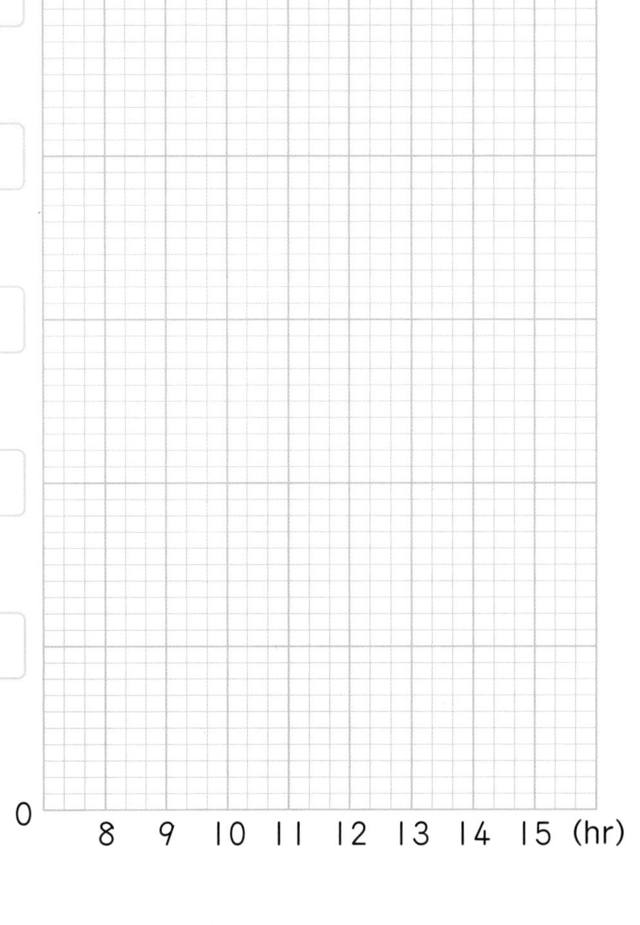

Changing the color and trace type of the two lines of the graph makes it easier for us to understand.

Why did the length of the shadow become the shortest at noon?

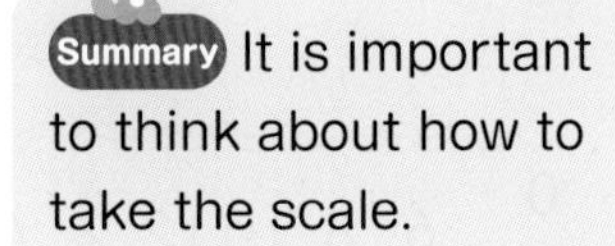

Summary It is important to think about how to take the scale.

We have fever when we have a cold.

I want to find out at what time I had high fever.

3 Ideas for drawing a line graph

1 **When Yukie had a cold, she investigated how her body temperature changed and represented it in a line graph. Let's think about this graph.**

1. What was her temperature (℃) at 8 a.m. and 2 p.m.?
2. Yukie redrew the following graph in order to understand better about how her body temperature changed. Let's discuss her new idea.

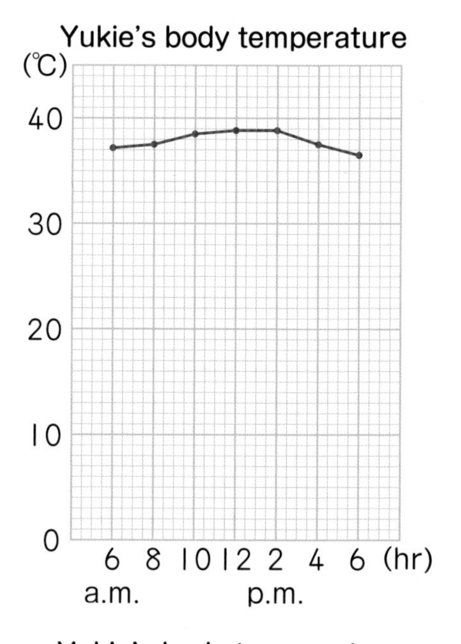

\ Want to think /

Purpose What are the differences between the two graphs?

How many grid lines did she use for 1℃?

What does 〰 mean?

Between what time did the body temperature change the most?

Which graph is easier to understand the change of her body temperature?

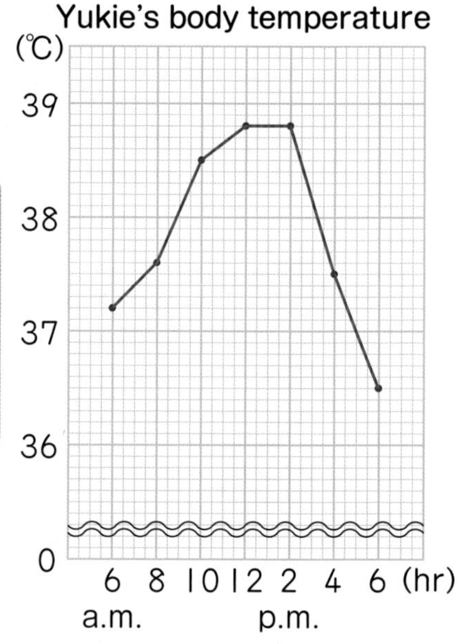

CAN What can you do?

☐ We understand the rate of change from graphs. → pp.28 ~ 30

1 The following table shows how the temperature changed at points near the window and the corridor in Yumi's classroom. Let's answer the following questions.

Temperature change in the window side and the corridor side

Time (hr)	a.m. 9	10	11	12	p.m. 1	2	3
Temperature near the window (℃)	15	18	20	21	23	23	22
Temperature near the corridor (℃)	15	16	18	18	20	20	19

① Let's represent how the temperature changed in a line graph.

② Between what time did the temperature change the most?

③ Between what time did not change?

④ Look at the graph and write down what you found out.

Supplementary Problems → p.143

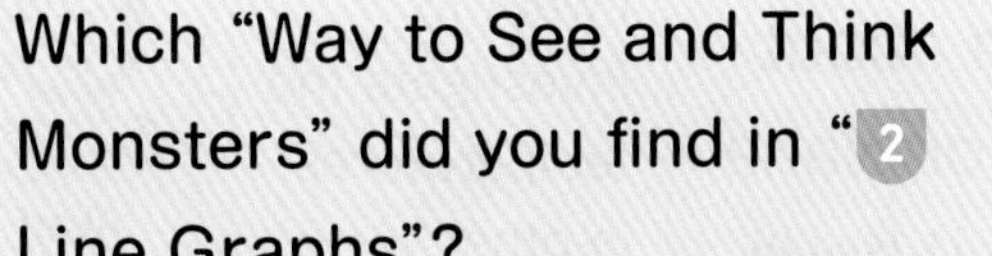

Which "Way to See and Think Monsters" did you find in "2 Line Graphs"?

I found "Other Way" when I was summarizing the results.

Akari

I found other monsters too!

Haruto

Usefulness and Efficiency of Learning

1 The table below shows how the temperature changed in Auckland City. Let's think about the following questions.

Average temperature per month in Auckland City

Month	1	2	3	4	5	6	7	8	9	10	11	12
Temperature (℃)	20	21	19	17	14	13	12	12	13	16	18	19

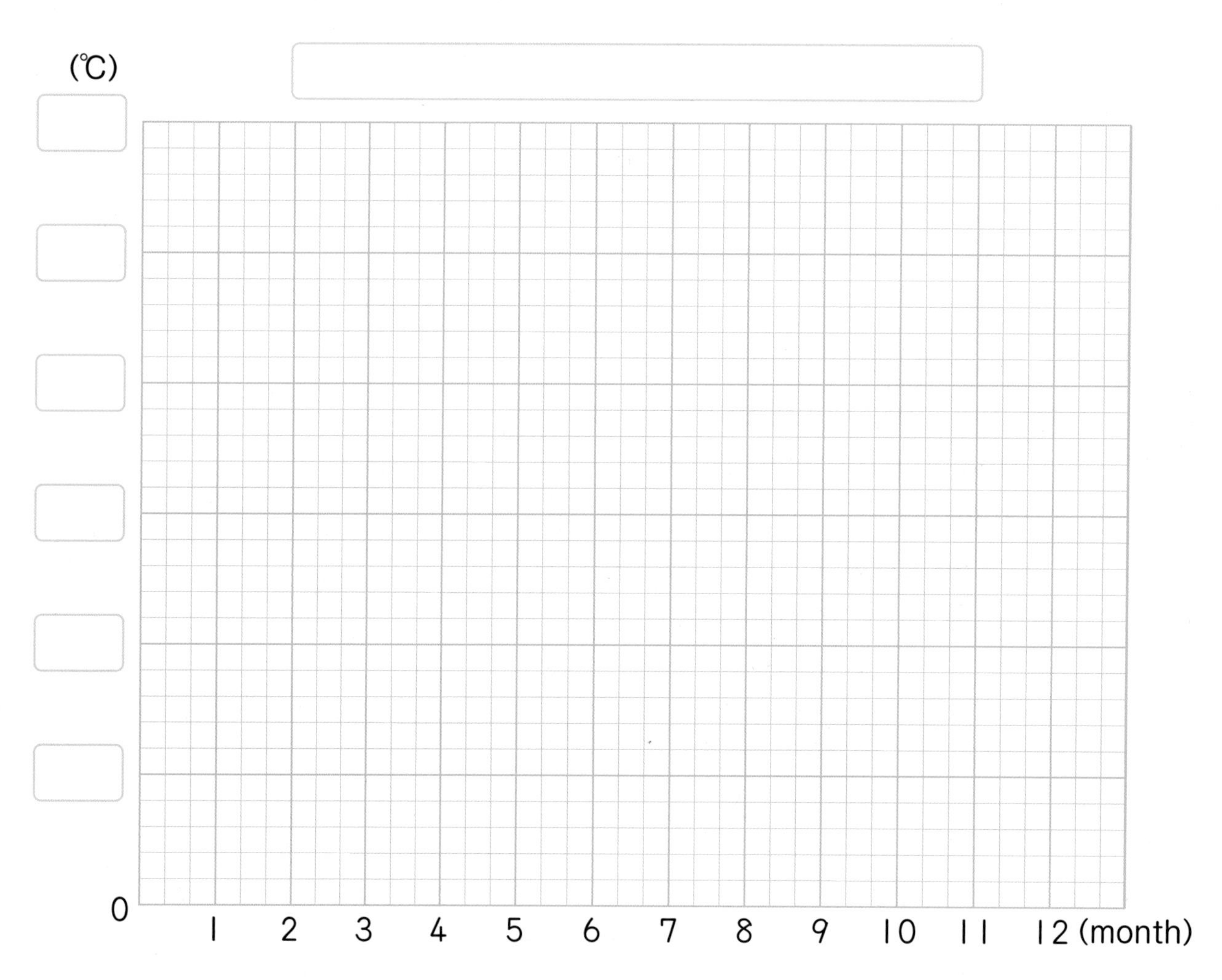

① Let's represent this information in a line graph.

② Let's compare this graph with the graph about the temperature changes in Niigata City or Naha City on page 28, and write down what you have noticed.

I want to investigate the temperatures in other countries.

With the Way to See and Think Monsters...

Let's Reflect!

Let's reflect on which monster you used while learning "2 Line Graphs."

Other Way

We could understand easily by representing the results in tables, bar graphs, or line graphs.

① Let's reflect on the various graphs we have learned.

Average temperature per month in Niigata City

Month	1	2	3	4	5	6	7	8	9	10	11	12
Temperature in Niigata City (℃)	5	5	8	10	17	22	24	28	24	16	11	5

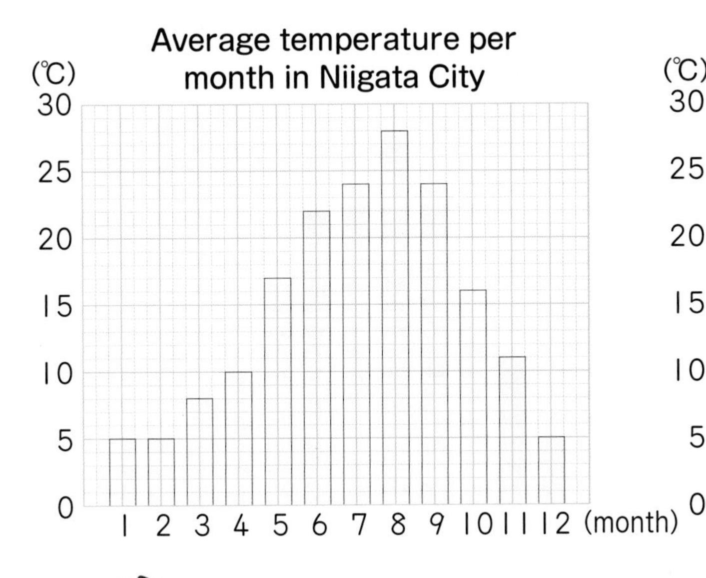

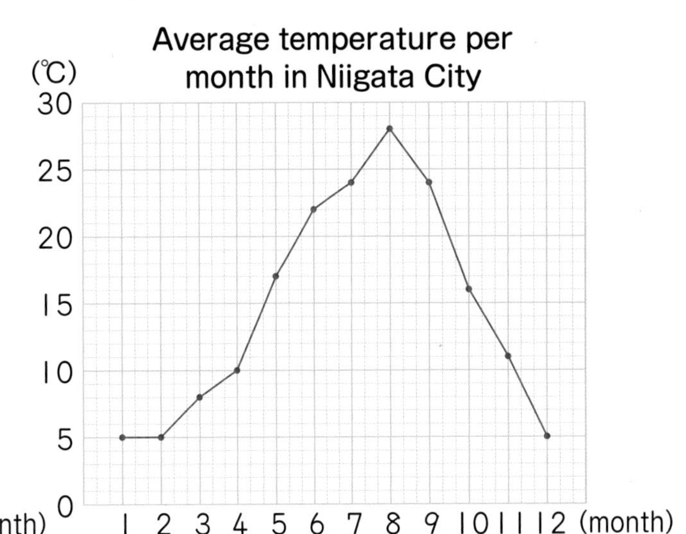

Haruto: Tables makes it easier to find out the temperature.

Sara: Bar graphs show the difference of the temperature, and line graphs show the change of the temperature.

Solve the ?

Line graphs makes it easier to see the change.

Want to Connect

Are there other graphs than bar graphs and line graphs?

Reflect Connect

Problem **Is there a difference between the amount of reading between two schools?**

> Let's write down what you can read from the bar graph and the table.

(Bar graph)

- As for school A, Story is the most borrowed type of books.
- As for school B, Story and History are the most borrowed type of books.
- As for school A and B, Picture Book has the same number of borrowed books.

(Table)

- In a period of 4 months, School A borrowed more books.

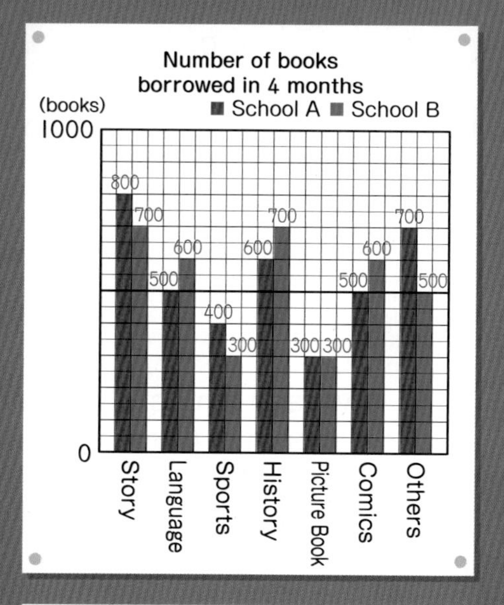

(books)

Type of books	School A	School B	Total
Story	800	700	1500
Language	500	600	1100
Sports	400	300	700
History	600	700	1300
Picture Book	300	300	600
Comics	500	600	1100
Others	700	500	1200
Total	3800	3700	7500

Haruto

In a bar graph, it is easy to understand which type of book was borrowed the most.

Sara

It is difficult to understand how much the increase is in a bar graph.

Yu

In a line graph, increases and decreases are easy to understand.

Let's read how the number of borrowed books changed from the line graph

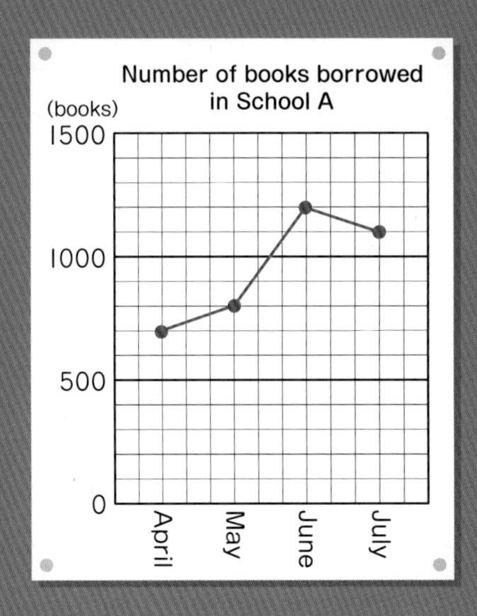

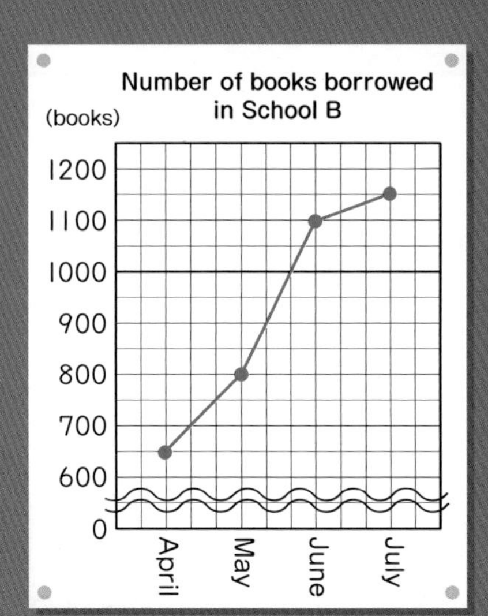

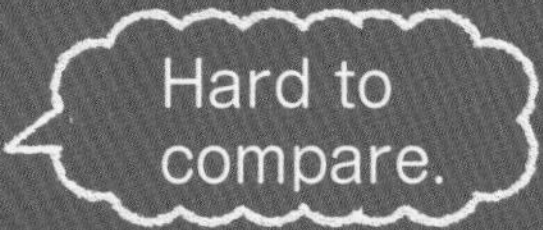

Need the same scale.

If the scale is aligned...

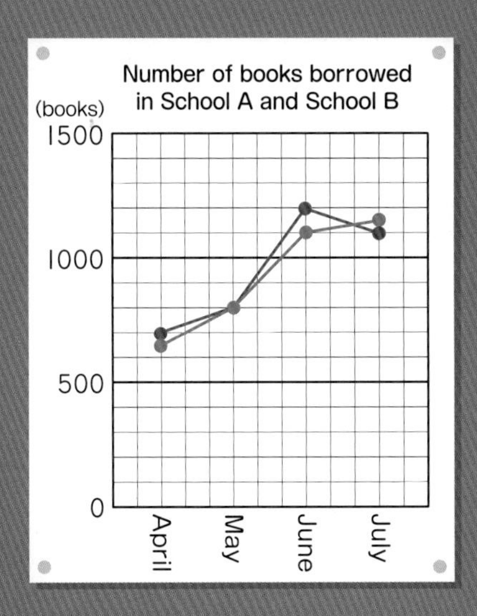

Things observed and understood from the line graph

- The highest increase was presented in School A between May and June.
- The amount of borrowed books in School A decreased between June and July.
- The amount of borrowed books in School B increased between April and July.

Changes are easy to understand in a line graph.

Akari

In line graphs, it is difficult to compare if they don't have the same scale.

Want to Connect

In this line graph, we cannot understand what type of books increased. I also want to create a line graph with type of books.

Haruto

Find the ?

Which one is the winning card?

Let's play lottery using division.

?	?
?	?
?	?

【Rule】
The winning card is the one which the quotient of the division expression written on the card is 3.

1

Winning cards	Losing cards
6 ÷ 2 =	12 ÷ 6 =
24 ÷ 8 =	12 ÷ 2 =
3 ÷ 1 =	12 ÷ 3 =

12÷4=

Look at this card. Is

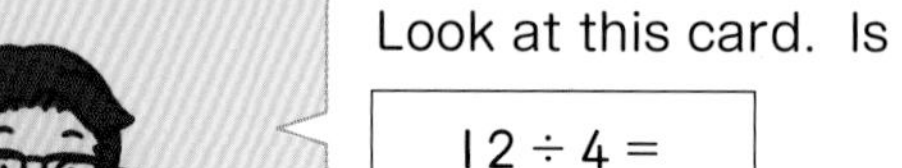

this card a winning card?

12 ÷ 4 = 3, so it's a winning card.

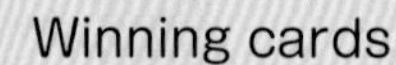

2

I sorted the winning cards.

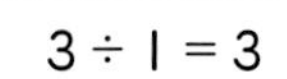

3 ÷ 1 = 3

6 ÷ 2 = 3

12 ÷ 4 = 3

24 ÷ 8 = 3

Seems that there is an interesting rule.

3

I arranged the cards of 12.

12 ÷ 6 = 2	12 ÷ 2 = 6
12 ÷ 3 = 4	12 ÷ 4 = 3

Division where the dividend is 12 seems to have other rules.

4

What rules can we find in division?

3 Division

Let's explore the rules which we have found.

1 Rule of division

Divisions with the same quotient →

1 There is a ribbon which is □ m long. When we cut it in units of ○ m, we got 3 ribbons. Let's answer the following questions.

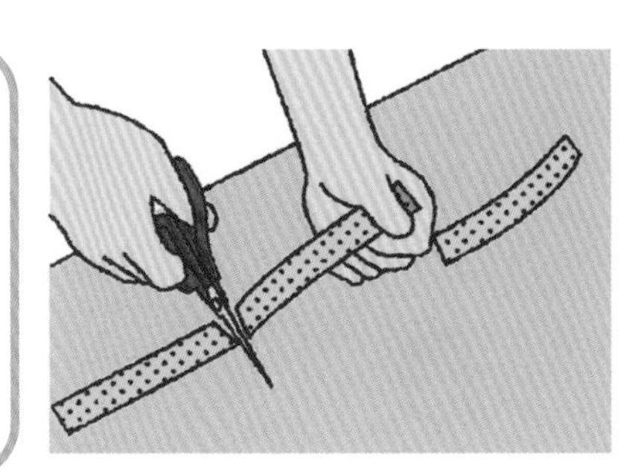

① Let's make a division expression using □ and ○.

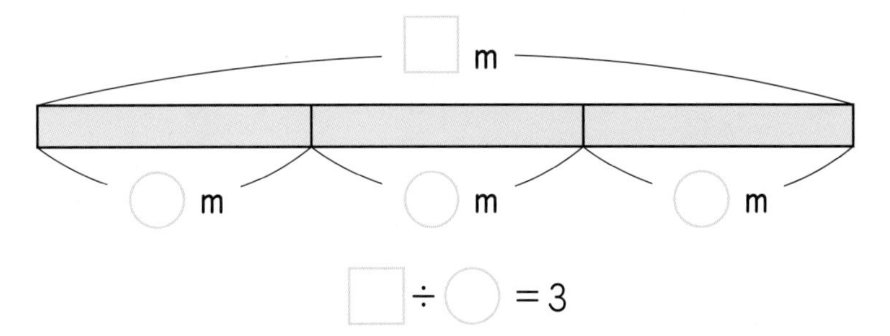

□ ÷ ○ = 3

② What number apply to □ and ○?

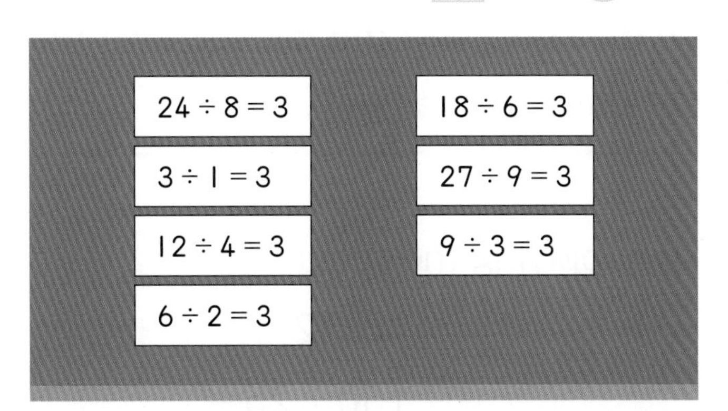

We can use the row of 3 of the multiplication table.

Haruto

? Want to find out: What rules can be found in the case of division expressions of same quotients?

③ Let's explain the rules that Akari and Yu found.

Akari's idea

$$3 \div 1 = 3 \xrightarrow{\times \square,\ \times \square} 6 \div 2 = 3 \xrightarrow{\times \square,\ \times \square} 12 \div 4 = 3 \xrightarrow{\times \square,\ \times \square} 24 \div 8 = 3$$

$$3 \div 1 = 3 \xrightarrow{\times \square,\ \times \square} 9 \div 3 = 3 \xrightarrow{\times \square,\ \times \square} 27 \div 9 = 3$$

Way to see and think

They are applying various numbers.

Yu's idea

$$12 \div 4 = 3 \xrightarrow{\div \square,\ \div \square} 6 \div 2 = 3 \xrightarrow{\div \square,\ \div \square} 3 \div 1 = 3$$

$$27 \div 9 = 3 \xrightarrow{\div \square,\ \div \square} 9 \div 3 = 3 \xrightarrow{\div \square,\ \div \square} 3 \div 1 = 3$$

Can we find out by using $18 \div 6 = 3$?

Sara

Summary

In division, even if we multiply the same number to the dividend and the divisor, the quotient doesn't change. Also, even if we divide the dividend and the divisor with the same number, the quotient doesn't change.

Way to see and think

1 Let's fill in each $\square$ by applying the rule above.

① $16 \div 8 = 4 \div \square$ ② $16 \div 2 = \square \div 6$

③ $270 \div 90 = 27 \div \square$ ④ $150 \div 25 = \square \div 100$

What happens to the quotient when we keep the dividend is the same, and change the divisor?

2

There are 12 candies. If we distribute □ to ○ people, we could divide the candies without any remainder. Let's answer the following questions.

① If you divide [2] candies to each, how many children can share the candies?　　$12 \div 2 =$ ○　　○ children

② Let's use □ and ○ and make a division sentence.

③ Let's find out the numbers for □ and ○.

? Purpose ＼ Want to find out ／ **What rules can be found between the divisor and the quotient?**

④ Let's explain the rule Yu found.

Yu's idea

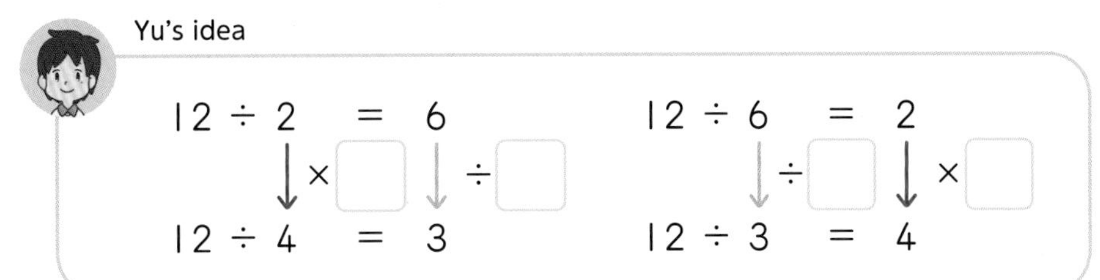

! Summary Way to see and think

In division, if you multiply the divisor □ times, the quotient will be the number which was divided by □. If you divide the divisor with □, the quotient will be □ times.

⑤ Let's continue Sara's idea on the right.

Sara's idea

When the divisor and the quotient…

$12 \div 2 = 6$　　$12 \div 3 = 4$

$12 \div 6 = 2$　　$12 \div 4 = 3$

1 When we distribute □ candies to ○ people, we could divide the candies without any remainder. Let's answer the following questions.

① Let's use □ and ○ and make a division sentence.

② Let's find out the numbers for □ and ○.

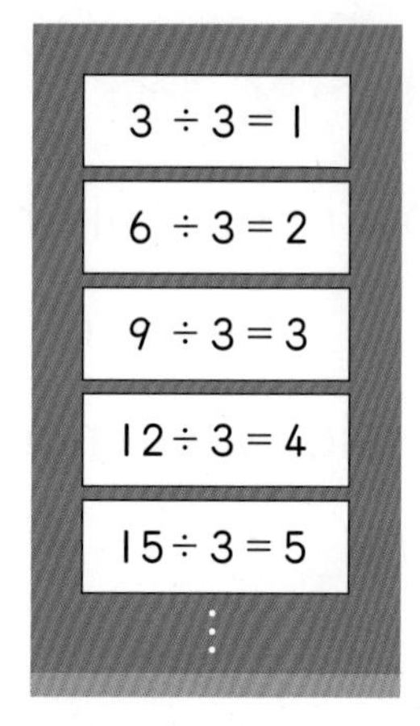

The dividend is the same as the answers for the row of 3 of the multiplication table.

I'm sure that the math sentence will continue after this...

③ Let's explain the rules Yu and Sara found.

Yu's idea

If the quotient increases by 1, the divisor increases by □.

+3 (3 ÷ 3 = 1) +1
+3 (6 ÷ 3 = 2) +1
+3 (9 ÷ 3 = 3) +1
12 ÷ 3 = 4
⋮

Sara's idea

If the dividend is doubled, the quotient...

6 ÷ 3 = 2
↓ × □ ↓ × □
12 ÷ 3 = 4

If the dividend is divided by 2, the quotient...

18 ÷ 3 = 6
↓ ÷ □ ↓ ÷ □
9 ÷ 3 = 3

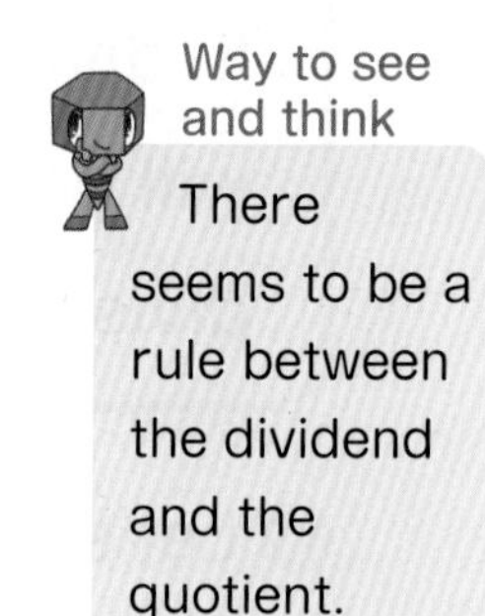

Way to see and think
There seems to be a rule between the dividend and the quotient.

2 Let's apply the rules we have found to other cases of division.

2 Division of tens and hundreds

1 There are 60 sheets of colored paper. We divide them equally among 3 children. How many sheets will one person get?

❶ Let's write a math expression. ☐ ÷ ☐

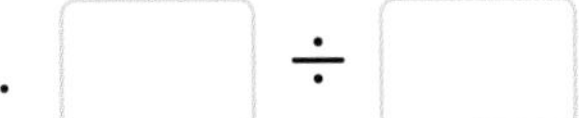

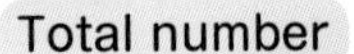

Total number　Number of children

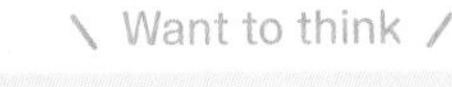

Purpose How can we think about this problem?

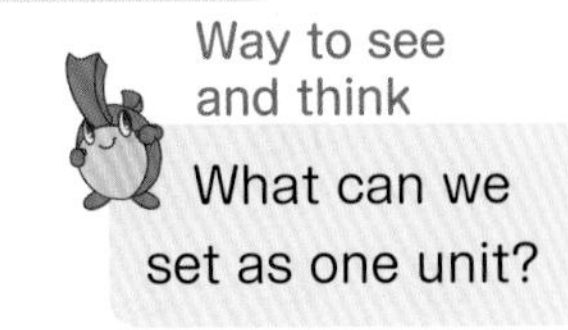

❷ Let's think about the number of 10-sheet bundles, and write a math expression.

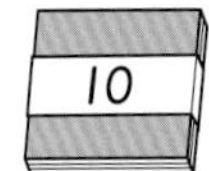

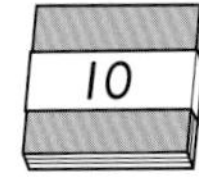

☐ ÷ ☐

Number of 10-sheet bundles　Number of children

❸ How many sheets will each child receive?

❹ Compare the two math expressions in ❶ and ❷, and discuss what you have noticed.

1▶ We want to divide 600 sheets of colored paper so that 3 children receive equally. How many sheets of paper will each child receive?

① Let's write a math expression.

② By setting how many sheets of bundles as one unit can we calculate using 6 ÷ 3?

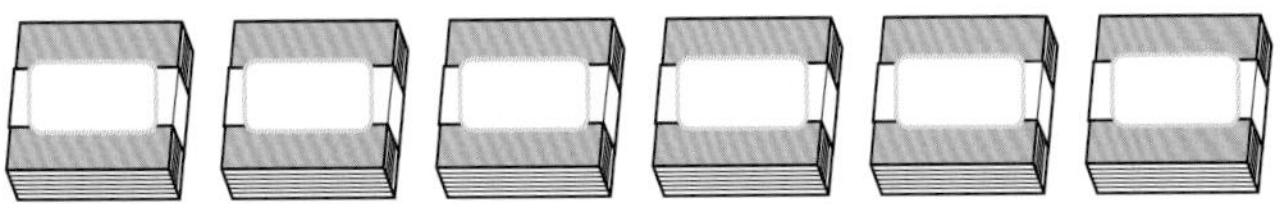

Summary We can calculate by setting 10 or 100 as one unit.

Yu

③ How many sheets of paper will each child receive?

2▶ Let's calculate the following.

① 80 ÷ 2	② 60 ÷ 2	③ 150 ÷ 5
④ 800 ÷ 2	⑤ 600 ÷ 2	⑥ 1500 ÷ 5

CAN What can you do?

☐ We understand the rules of division. → pp.39 ~ 42

1 Let's fill in each ☐ based on the rules of division.

① $12 \div 2 = \square \div 4$

② $9 \div 3 = 18 \div \square$

③ $18 \div 2 = 9$
↓ ×3 ↓ ÷☐
$18 \div 6 = 3$

④ $24 \div 6 = 4$
↓ ÷2 ↓ ×☐
$24 \div 3 = \square$

⑤ $4 \div 2 = 2$
↓ ×☐ ↓ ×☐
$40 \div 2 = \square$

⑥ $16 \div 2 = 8$
↓ ÷2 ↓ ÷☐
$8 \div 2 = 4$

☐ We can calculate divisions of tens and hundreds. → p.43

2 Let's calculate the following.

① $40 \div 4$ ② $80 \div 4$ ③ $90 \div 3$

④ $400 \div 4$ ⑤ $800 \div 4$ ⑥ $900 \div 3$

☐ We can calculate division in large numbers using the rules of division. → p.43

3 We want to divide 1200 sheets of colored paper equally among 3 children. How many sheets will each child receive?

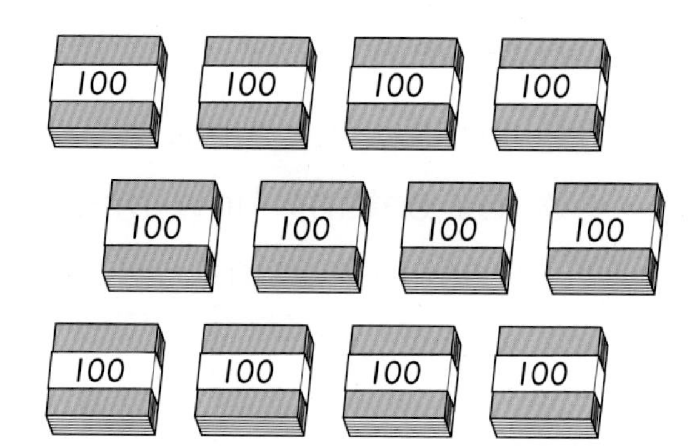

How many sheets of bundles can be set as one unit?

Supplementary Problems → p.144

Which "Way to See and Think Monsters" did you find in "3 Division"?

I found "Rule" when I was dealing with various divisions.

Haruto

When I was trying out a new idea for calculation...

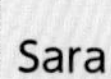

Sara

With the Way to See and Think Monsters...

Let's Reflect!

Let's reflect on which monster you used while learning "3 Division."

Rule

As for the relationship of the math sentences, we could find a rule by focusing on the quotient.

① What rules of division did you find out?

$9 \div 3 = 3$ (↓×3, ↓×3) $27 \div 9 = 3$

The quotient remains the same even if the dividend and the divisor are multiplied by the same number.

$27 \div 9 = 3$ (↓÷3, ↓÷3) $9 \div 3 = 3$

The quotient remains the same even if the dividend and divisor are divided by the same number.

Haruto

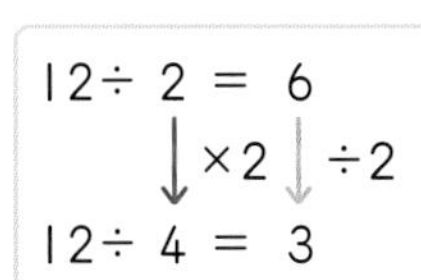

If you multiply the divisor by □, the quotient will be the number divided by □.

$12 \div 6 = 2$ (↓÷2, ↓×2) $12 \div 3 = 4$

If you divide the divisor by □, the quotient will be □ times.

Sara

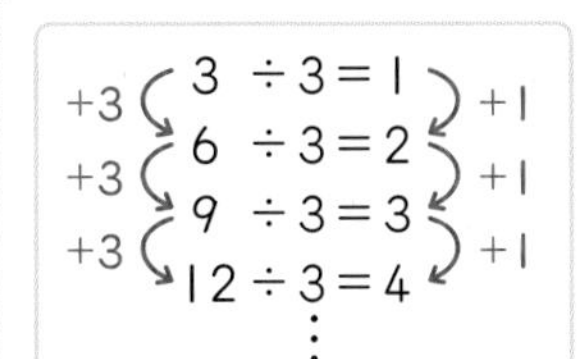

I found rules about the relationship between the dividend and the answer.

Yu

What rule can we find out from here?

$6 \div 3 = 2$ (↓×□, ↓×□) $12 \div 3 = 4$

$18 \div 3 = 6$ (↓÷□, ↓÷□) $9 \div 3 = 3$

Akari

Unit

By setting 10 and 100 as one unit, we could calculate large numbers.

② How could you calculate $120 \div 3$ and $1600 \div 4$ in easier ways?

By thinking in unit of □ ...

Haruto

? Solve the ?

We can calculate various divisions by using the rules of division.

Yu

Want to Connect

Can we calculate division with numbers beyond the multiplication table?

Akari

Find the ?

Which animal opens its mouth the widest?

Ⓐ

I think the hippo opens the largest.

Akari

Ⓒ

Ⓑ

We can compare the size of the angle.

Haruto

Ⓓ

Ⓔ

How can we compare?

Sara

\ Want to compare /

Purpose What can we do to compare the size of the angles?

Angles

Let's think about how to measure and draw angles.

1 The size of angles

How to compare the size of angles ↓

1 The animals Ⓐ, Ⓑ, Ⓒ, Ⓓ, Ⓔ are opening their mouth. Let's explore the size of the angles of their mouth.

1. Which animal opens its mouth the widest?
2. Which animal opens its mouth the narrowest?
3. Let's name the angles in order of the size of the angle from the widest to the narrowest.

Yu's idea

I will trace an angle on a transparent paper and compare them by placing one over the other.

Sara's idea

I will examine by how many times the size of an angle is using the triangle ruler.

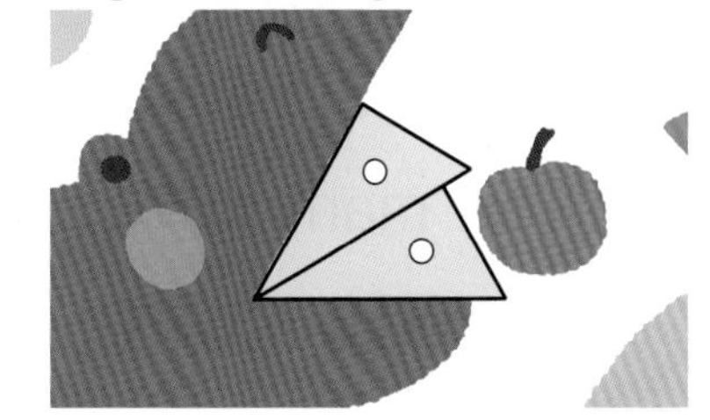

Summary

The size of an angle is not related to the length of the sides, but is determined by the opening between the sides. We can compare by directly comparing the size or by using a triangle ruler.

? Can we compare any size of angles?

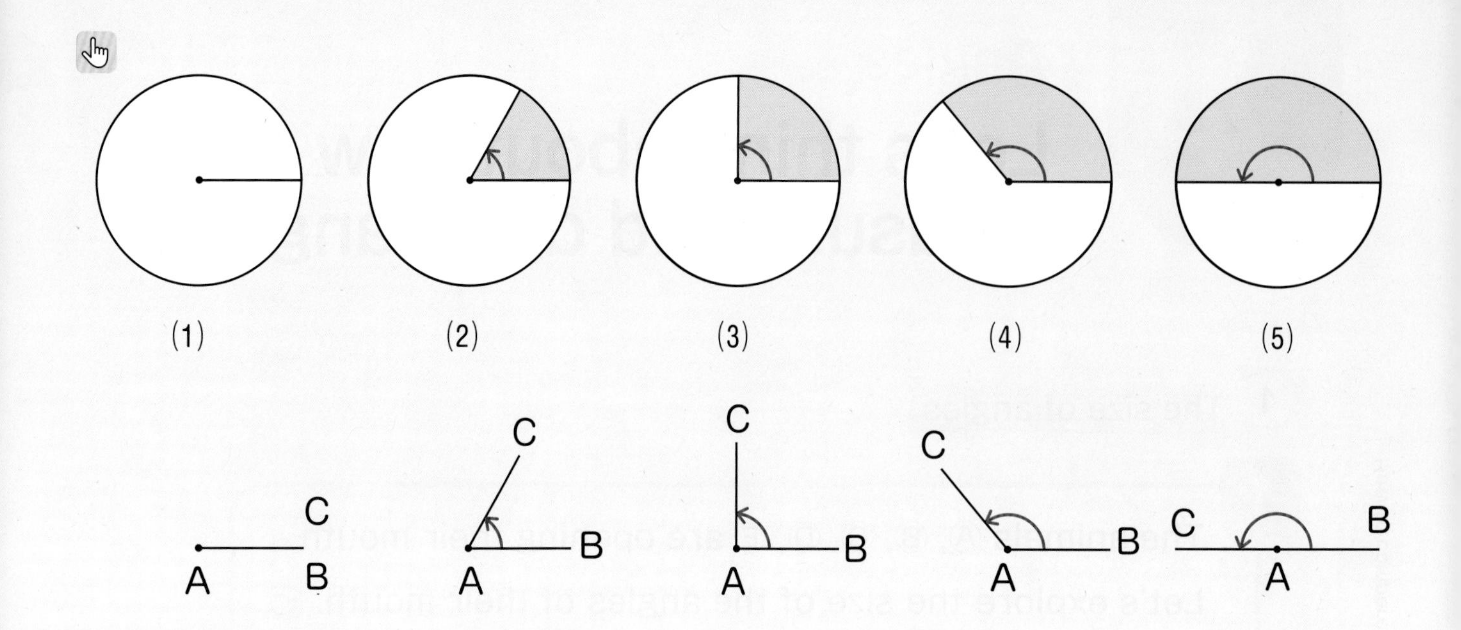

2 The size of rotating angles

The size of rotating angles ↓

1 Let's explore the rotated angle shown on the diagram above.

① Centered on Point A, if side AC rotates on the direction of the arrow, how does the size of the angle change?

② As for the angles (1) to (9), which one will become a right angle?

③ How many right angle parts will make the size of angles of (5), (7), and (9)?

The size of angle (5) is made by 2 right angle parts, so it is called 2 right angles. Angles (3), (7), and (9) are called 1 right angle, 3 right angles, and 4 right angles respectively.

4 right angles are called "angle of one revolution" and 2 right angles are called "angle of half revolution."

Can we represent the size of the angle for (1) to (9) by setting the angle of a triangle ruler as a unit?

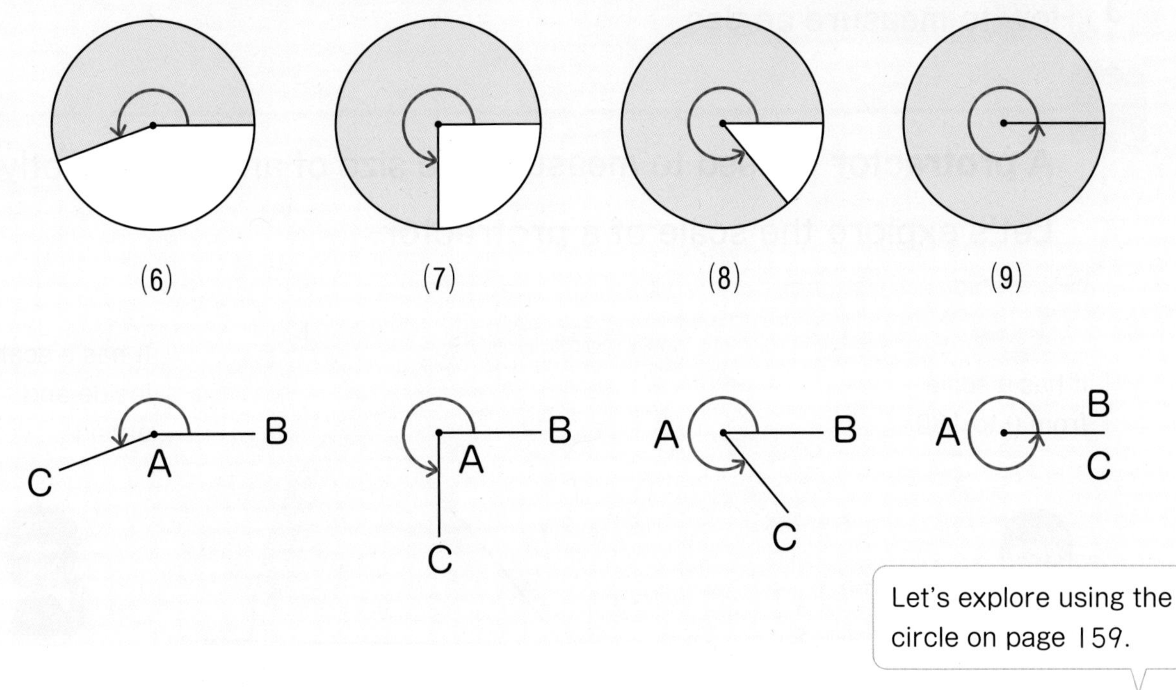

Let's explore using the circle on page 159.

④ Let's use a triangle ruler to make the size of angles (2) and (3).

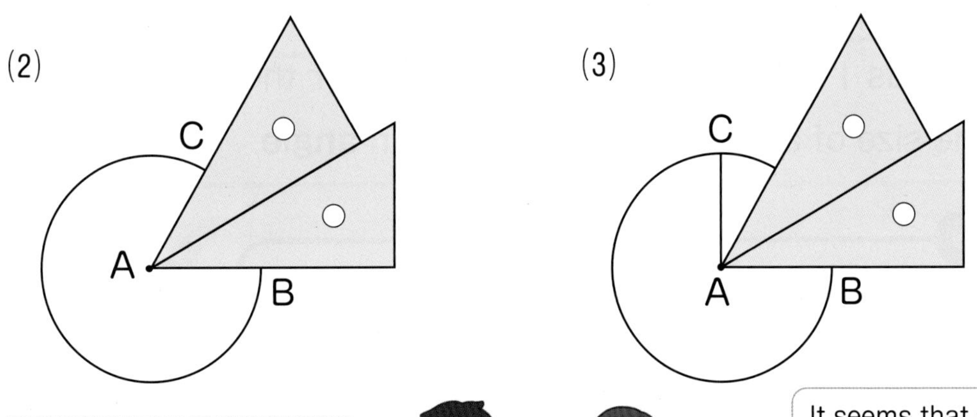

The angle (2) is two angles of the triangle ruler.
Can we examine the size of other angles?
It seems that one more angle can fit in angle (3).
It is easy if we can represent the size of angles like we did in length and amount of water.

\ Want to know /

? **Purpose** **Can we represent the size of an angle with a number?**

3 How to measure angles

1

A **protractor** is used to measure the size of angles correctly.
Let's explore the scale of a protractor.

It has a scale from 0 to 180.

Akari

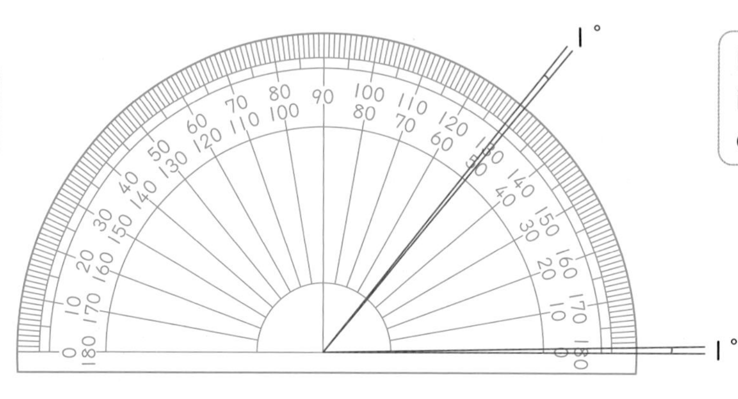

It has a scale inside and outside.

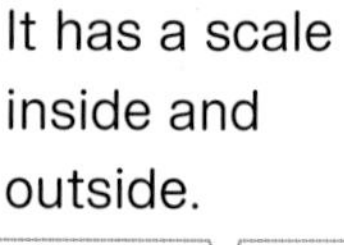

Haruto

One angle revolution has 360 equal parts. The size of an angle with one of those equal parts is written as 1° and is read as 1 **degree**. Degree is the unit for the size of an angle. The size of an angle is also called an **angle**.

Summary

As for the size of an angle, whatever part can be represented by using the 1° as one unit.

Way to see and think
We can represent angle in numbers by setting 1° as one unit.

1. What are the size of the angles (5), (7), and (9) on page 48, 49?

1 right angle = 90°

(5) 2 right angles = ☐°

(7) 3 right angles = ☐°

(9) 4 right angles = ☐°

(5)

(7)

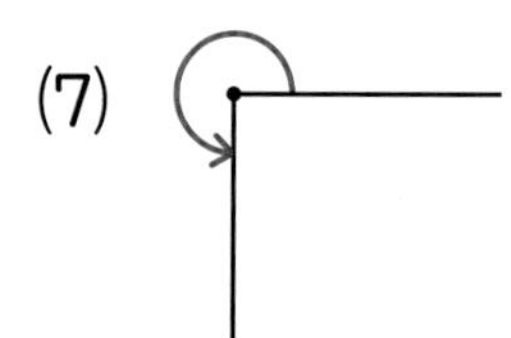

(9)

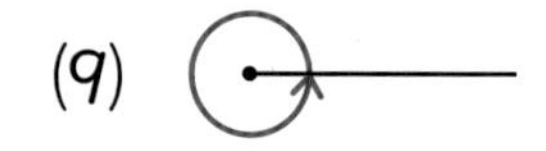

2 Let's measure the size of angle Ⓐ using the protractor.

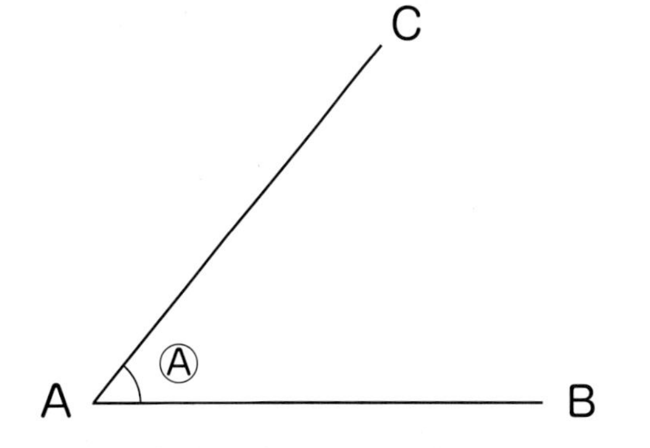

How to use a protractor

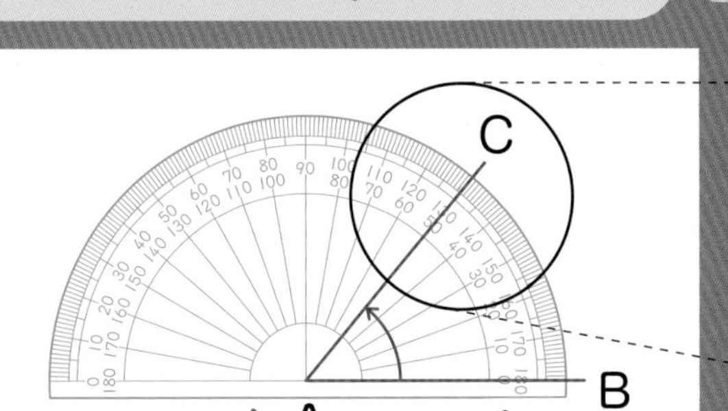

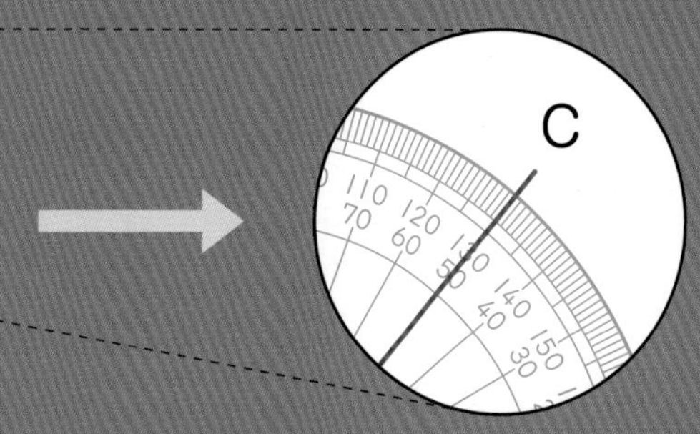

(1) Place the center of the protractor on vertex A.

(2) Place the 0° line on top of side AB.

(3) Read the scale that overlaps with side AC.

The size of angle Ⓐ is

 °.

Make sure not to measure in a wrong way.

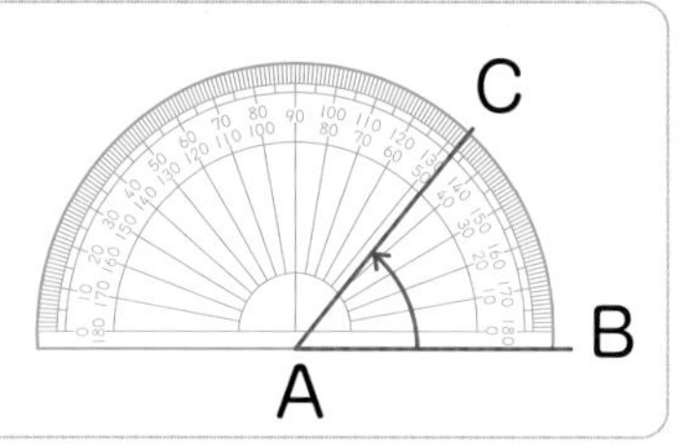

3 Let's measure the size of the following angles.

① 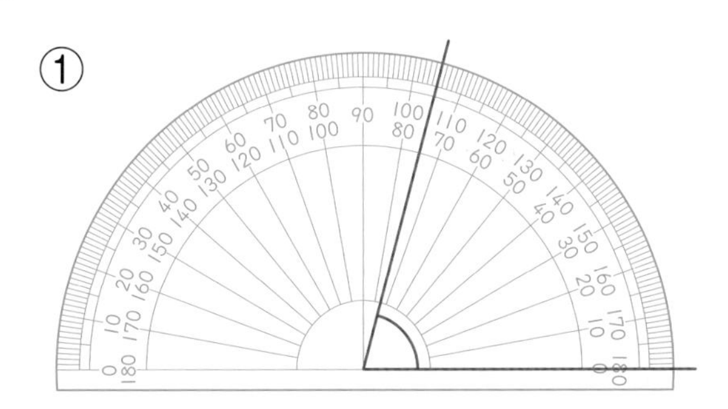

② 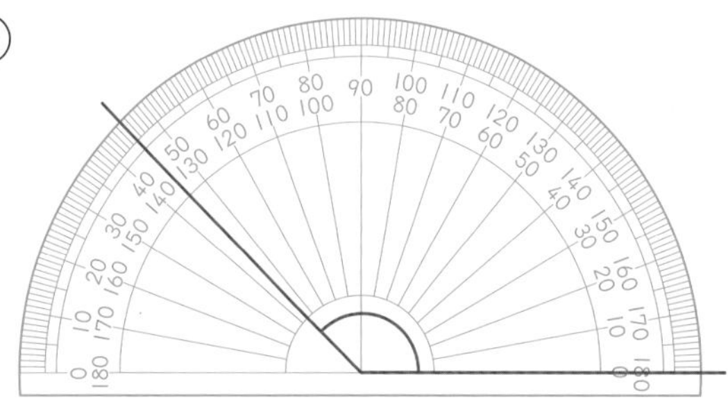

? Can we measure any size of angles by using a protractor?

2 Let's think about how to measure the following angles.

❶

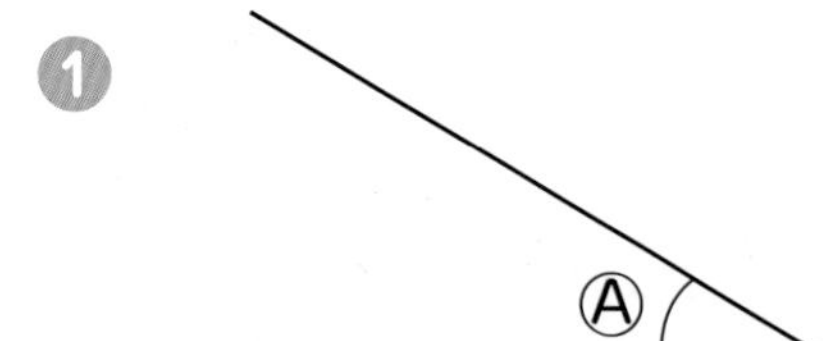

❷

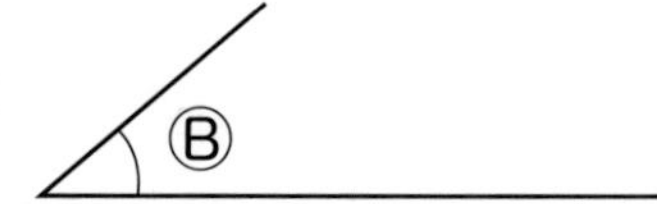

\ Want to try /

Purpose Can we measure various angles correctly?

Haruto

Sara's idea

Since the direction is opposite, I read the outside scale of the protractor.

Haruto's idea

Since the length of the side is short, I made it longer by extending it.

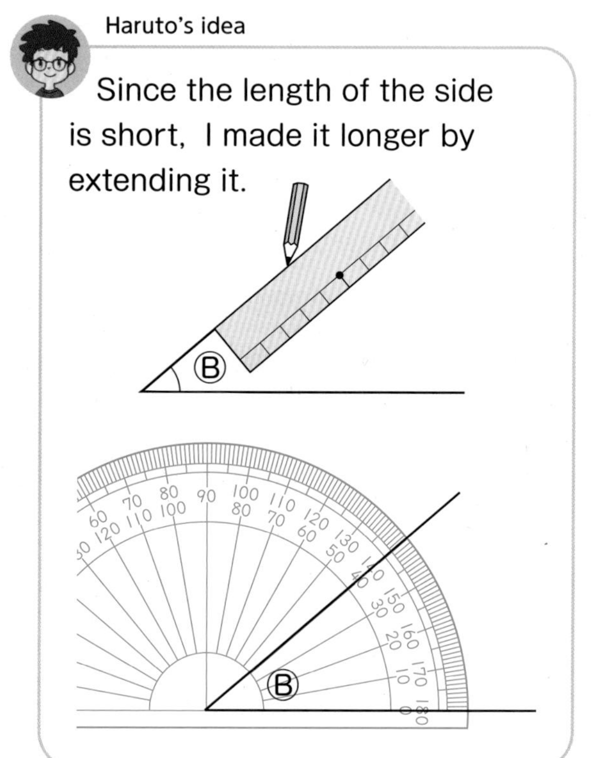

1 Let's measure the size of the following angles.

①

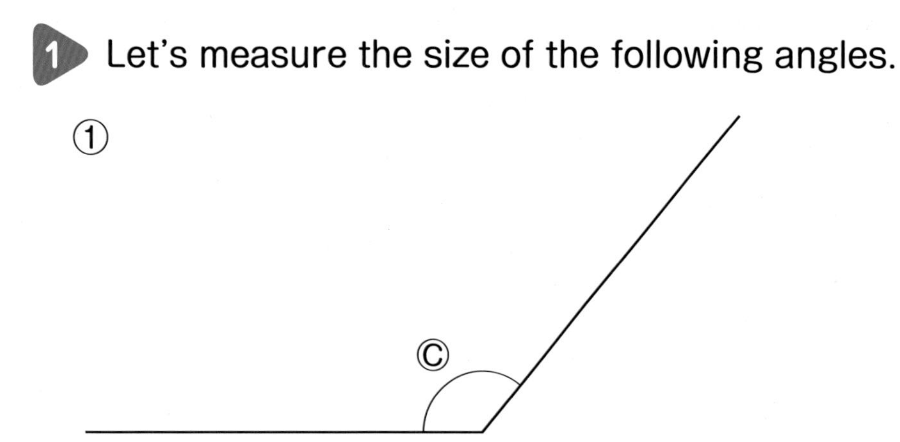

②

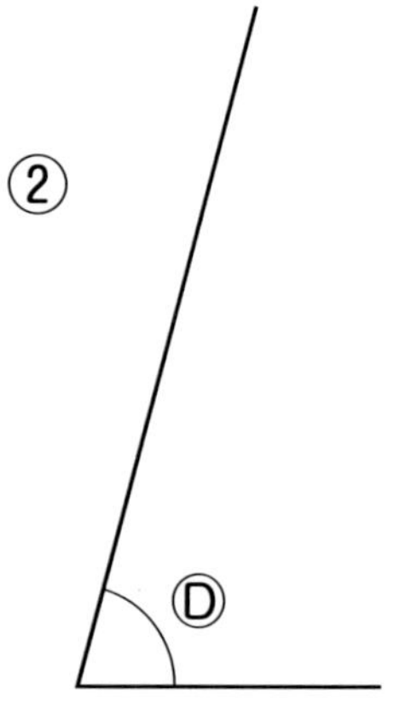

2 Let's measure the following angles.

Summary Before measuring, we need to check whether it is larger than a right angle, where to set the 0° line.

Akari

Ⓔ Ⓕ Ⓖ Ⓗ Ⓘ Ⓙ Ⓚ

3 Let's measure the angles of the mouth of the animals on page 46.

? Can I measure angles larger than 180°?

3 Let's measure angle Ⓐ with a new idea.

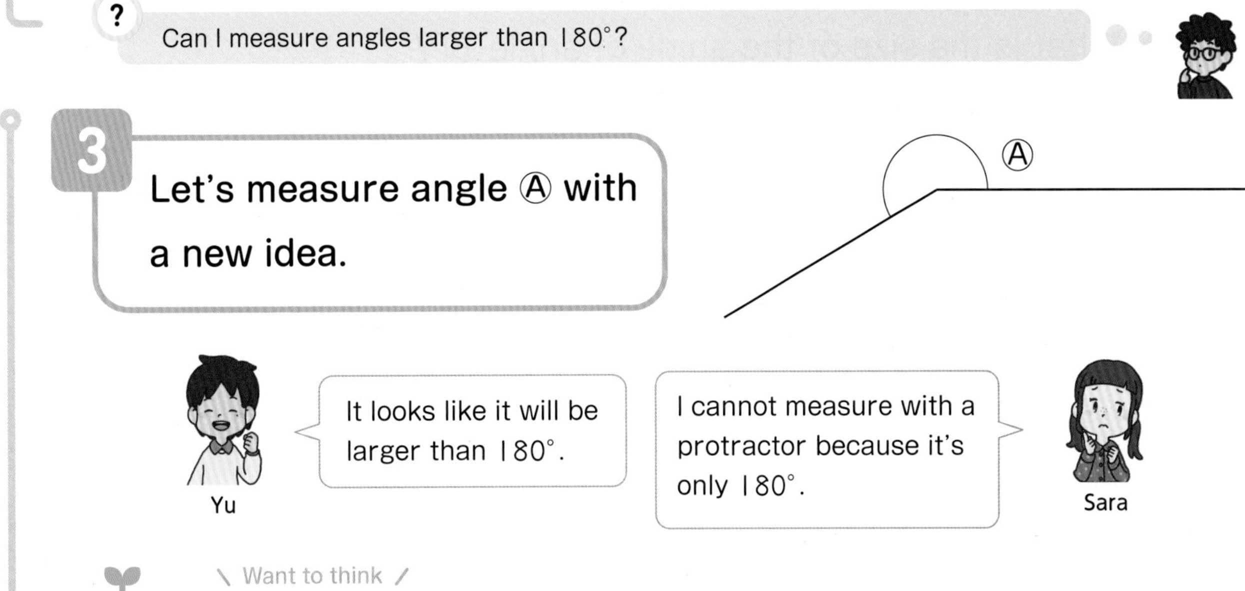

Ⓐ
It looks like it will be larger than 180°.
Yu
I cannot measure with a protractor because it's only 180°.
Sara

\ Want to think /

? **Purpose** How can we measure an angle larger than 180°?

① Let's discuss the new ideas of Haruto and Akari.

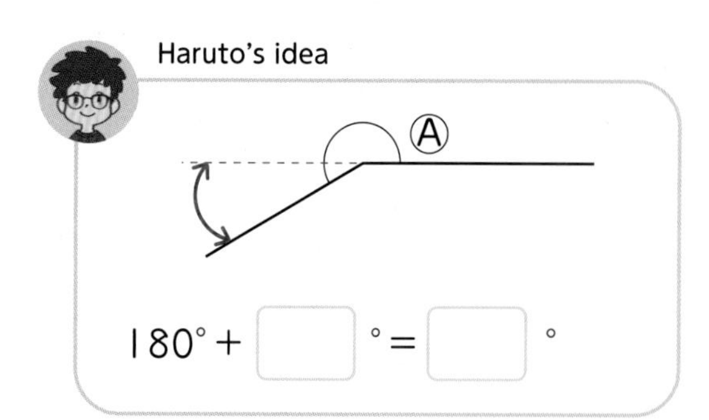

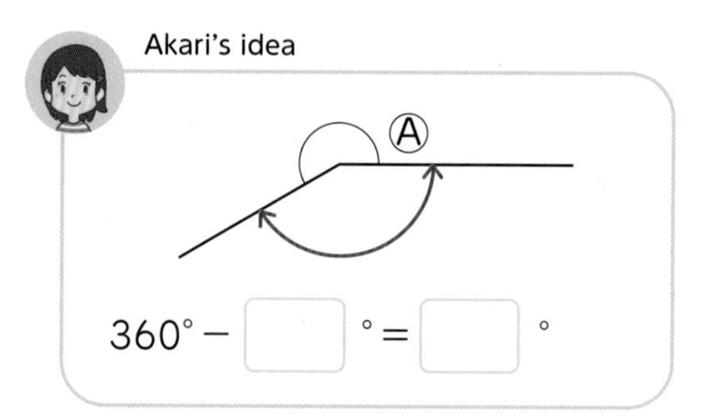

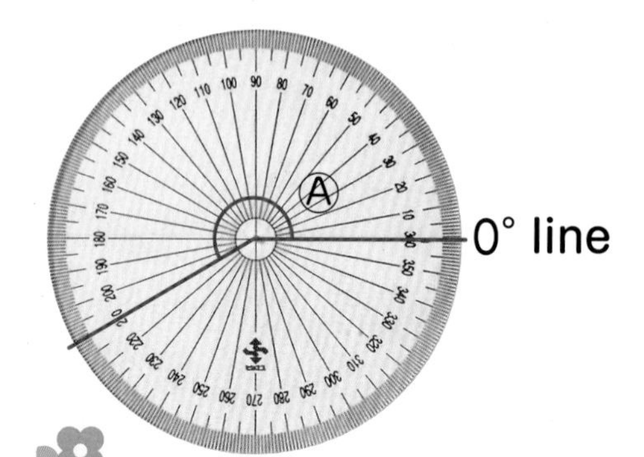

If you use an all round protractor, you can measure it at once.

Way to see and think

We can add or subtract the size of angles like we did in length or weight.

Summary

When you measure an angle larger than 180°, think about how many degrees larger it is than 180°, or how many degrees smaller it is than 360°.

Way to see and think

1 What is the size of the angle of angle Ⓑ?

2 The diagram on the right shows the angle formation by the intersection of two straight lines. Let's explore about this.

① When angle Ⓒ is 60°, what is the size of angle Ⓓ?

② Let's compare the sizes of angle Ⓒ and angle Ⓔ.

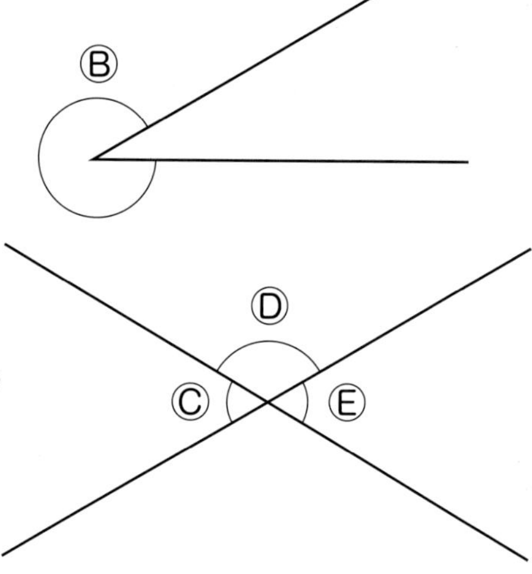

? Can we draw various angles using a protractor?

4 How to draw angles

1 Let's draw an angle with size 50° using a protractor.

\ Want to know /

Purpose How should I use the protractor?

How to draw angles ▷

(1) Draw one side AB.

(2) Place the center of the protractor and point A together. Align the 0° line with side AB.

(3) Place point C in the 50° scale place.

(4) Draw a straight line through points A and C.

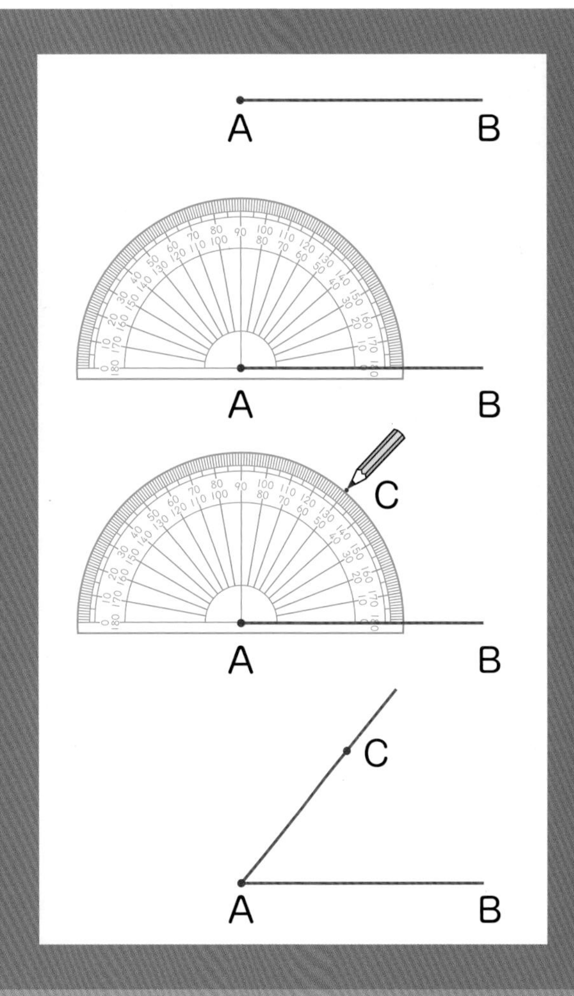

1 Let's draw angles with size 85° and 140°.

2 Let's draw an angle with size 210° using the ideas you have learned.

Can I use the same way to measure angles larger than 180°?

3 Let's draw the triangle shown on the right.

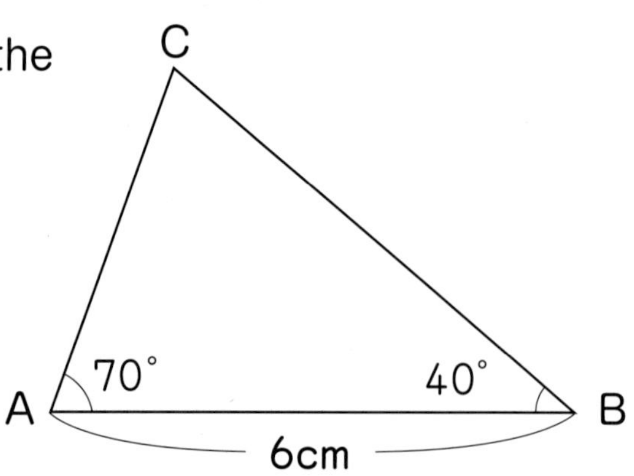

Where is the best place to start drawing?

After step (1), it's better to use a protractor.

How to draw triangles

(1) Draw side AB with the length of 6cm.

(2) Place the protractor center and point A together. Draw a 70° angle.

(3) Place the protractor center and point B together. Draw a 40° angle. Point C is the intersection point.

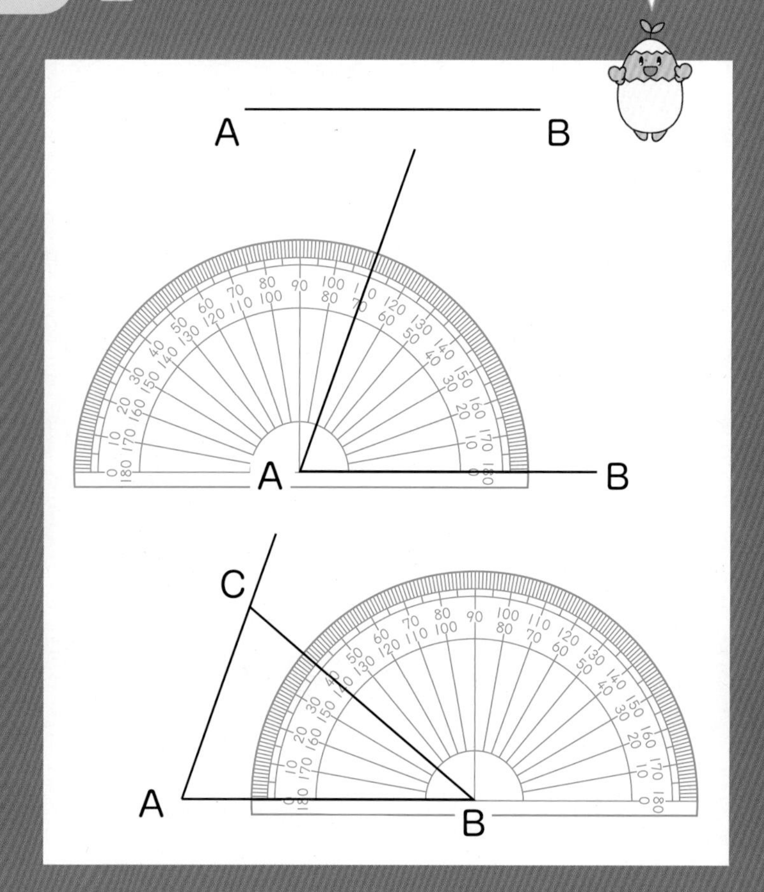

Summary By setting the center of the protractor on the point of the angle you want to draw, we can draw various angles and triangles.

4 Let's draw the following triangles.

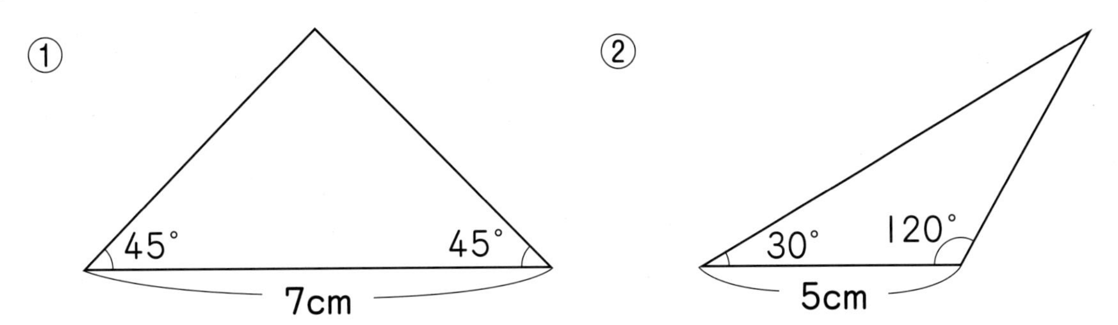

? I used a triangle ruler to measure the size of the angles in page 47. What are the sizes of the three angles of the triangle ruler?

That's it!

Making patterns using a protractor

① Let's divide the circle in order of 10°, 20°, 30°, ... until it overlaps on the start line.

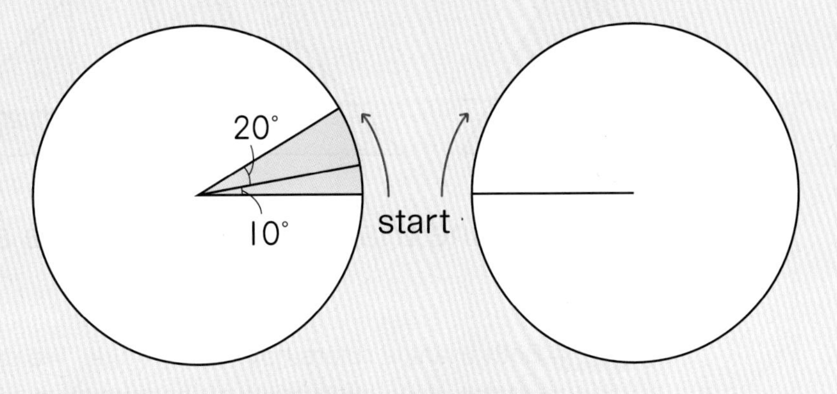

② Go 6cm to the right from the starting point, then make a 45° turn to the left. Then go another 6cm and make a 45° turn to the left. Let's do this repeatedly.

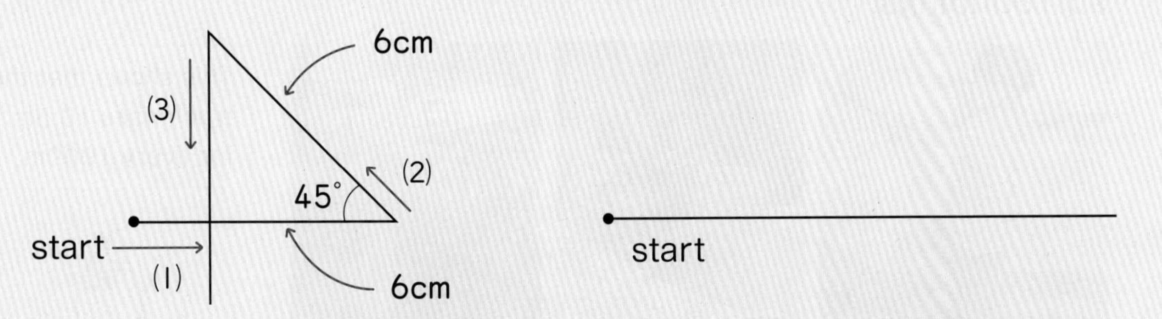

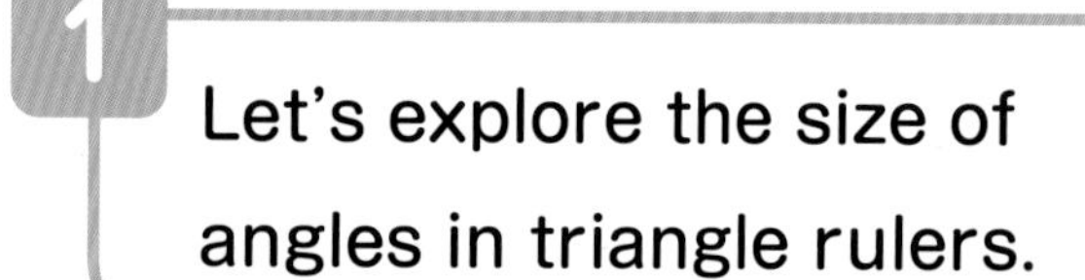

5 Angles of triangle rulers

1 Let's explore the size of angles in triangle rulers.

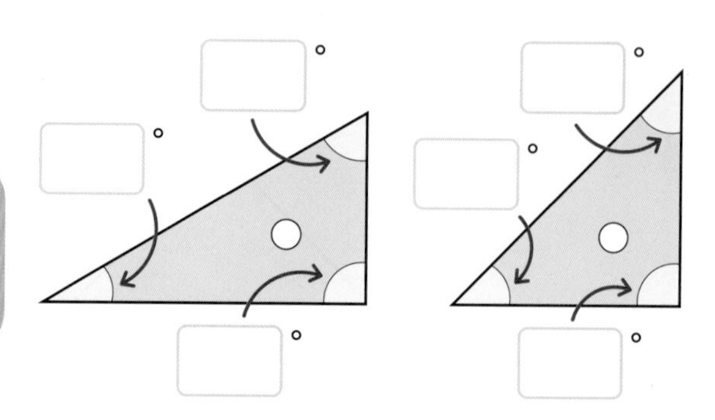

① Let's measure the size of each angle in one set of triangle rulers.

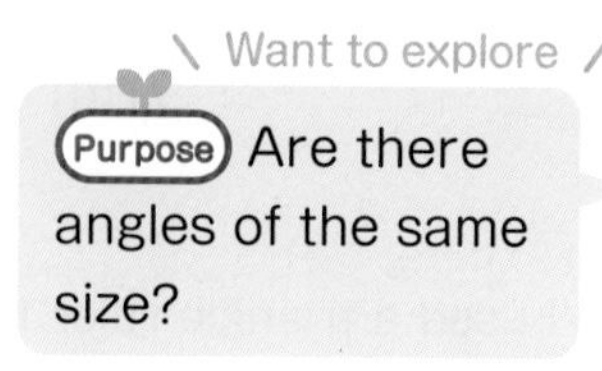

② Triangle rulers were combined to make angles as shown below. What are the sizes of each angle?

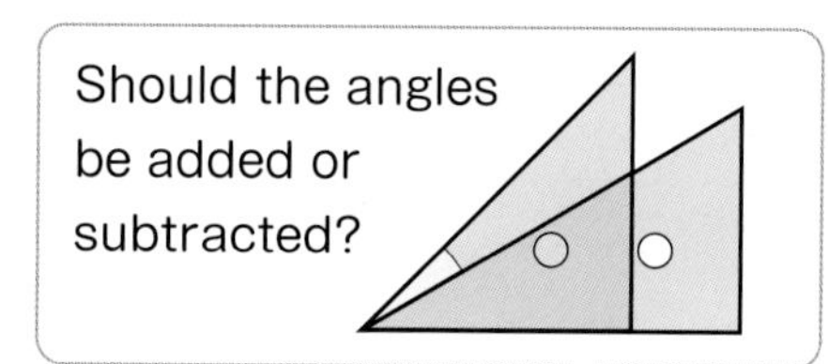

Should the angles be added or subtracted?

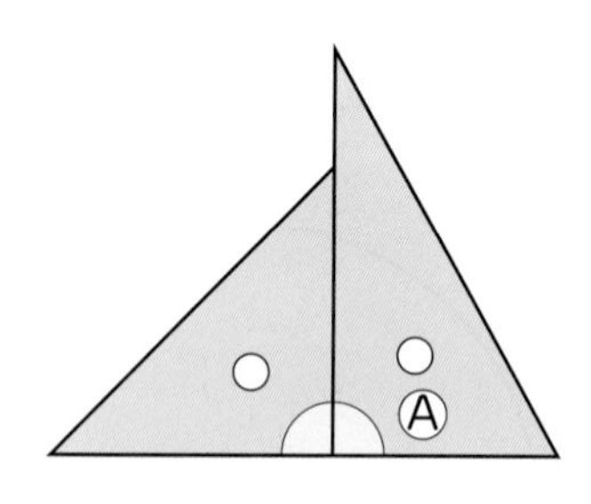

Ⓐ

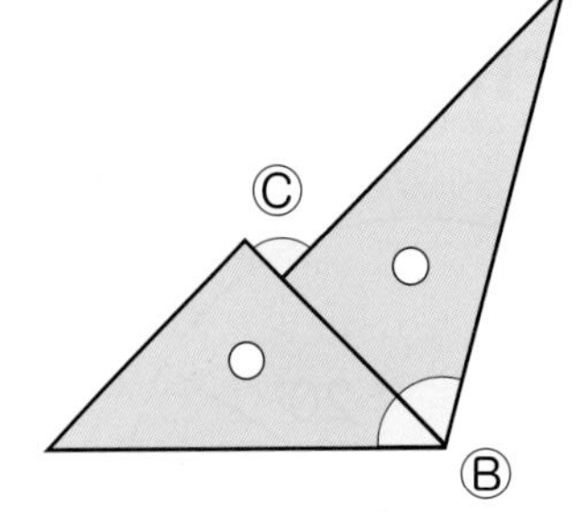

Ⓒ
Ⓑ

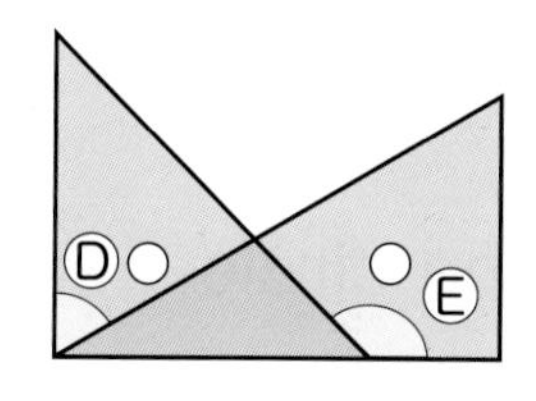

Ⓓ
Ⓔ

Sara

③ Let's combine triangle rulers to make a 150° angle.

Summary Both triangle rulers have a 90° angle. We can measure various angles using each angle of the triangle rulers.

1 Let's investigate angles in various places from your surroundings.

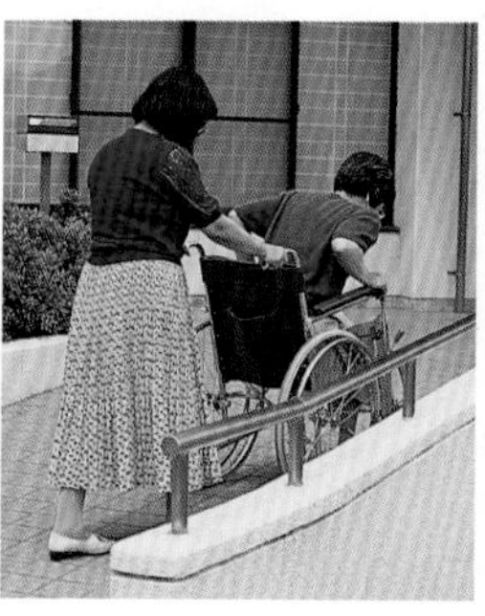

This shows that the track rises 10.5m for every 1000m.

CAN What can you do?

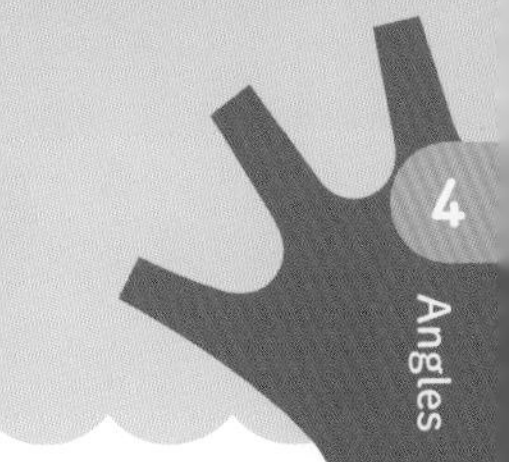

☐ We understand how to represent the size of an angle. → p.50

1 Let's summarize the size of an angle.

① The unit to represent the size of an angle is ☐ .

② One revolution angle has ☐ equal parts. The size of an angle with one single part is 1 degree.

☐ We can measure the size of an angle using the protractor. → pp.51 ~ 54

2 Let's measure the size of the following angles.

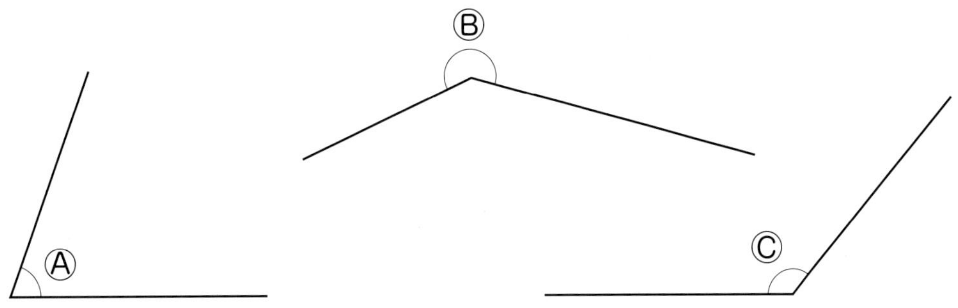

☐ We can draw an angle using the protractor. → p.55

3 Let's draw the angles with the size of 100° and 270°.

☐ We understand the angles of the triangle ruler. → p.58

4 Triangle rulers were combined to make angles as shown below. What are the sizes of angles Ⓓ, Ⓔ, and Ⓕ?

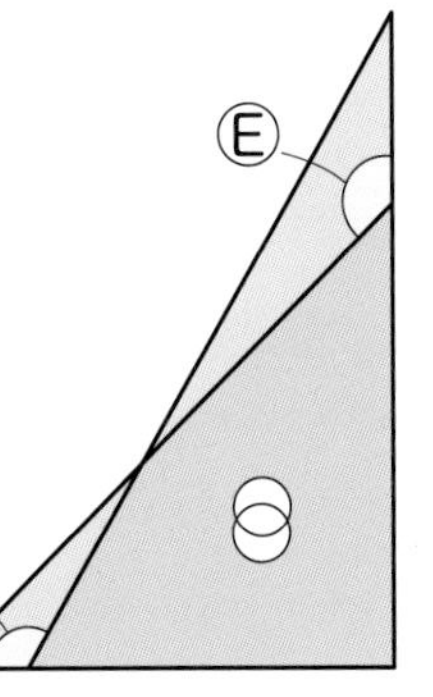

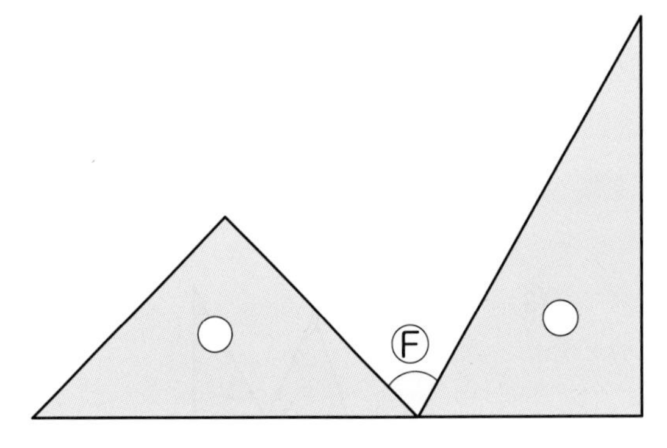

Supplementary Problems → p.145

Which "Way to See and Think Monsters" did you find in "4 Angles"?

I found "Unit" when I was trying to represent the size of angles in numbers.
Sara
When I was measuring the large size of an angle...
Yu

Usefulness and Efficiency of Learning

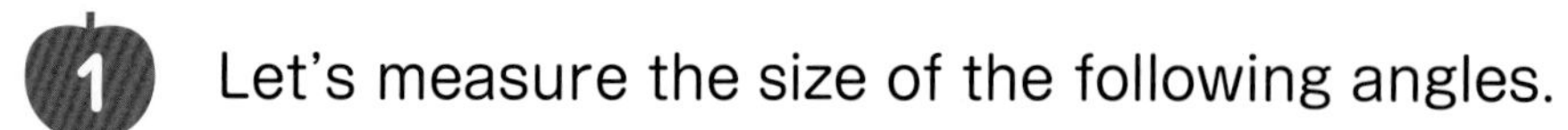

1 Let's measure the size of the following angles.

①

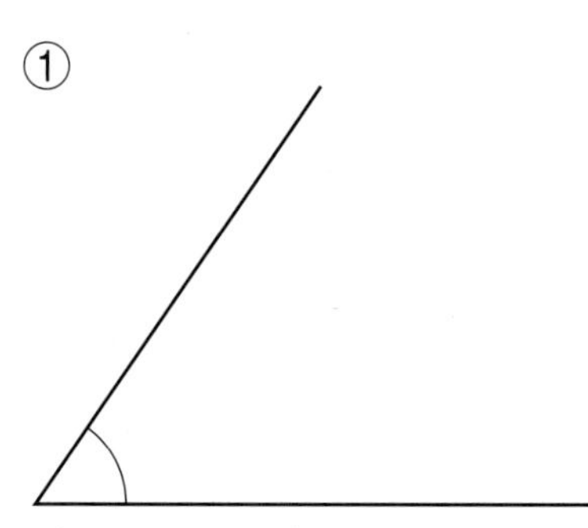

②

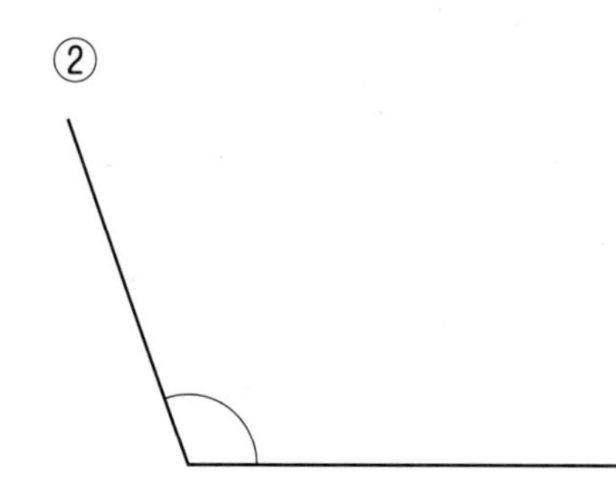

③

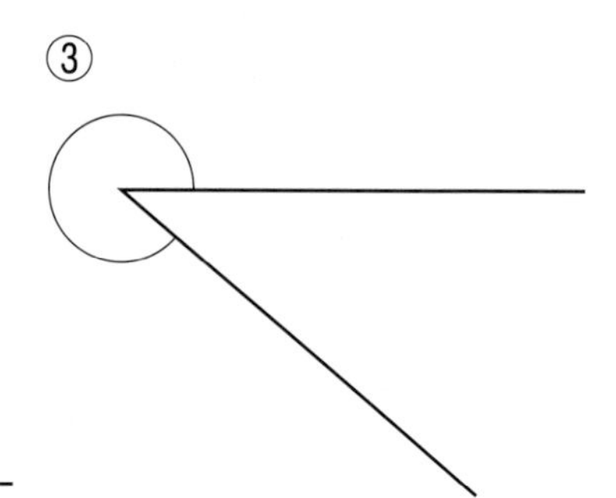

2 Let's draw angles of the following size.

① 120°　　　② 300°

3 Triangle rulers were combined to make angles as shown below. What are the sizes of angles Ⓐ to Ⓓ?

①

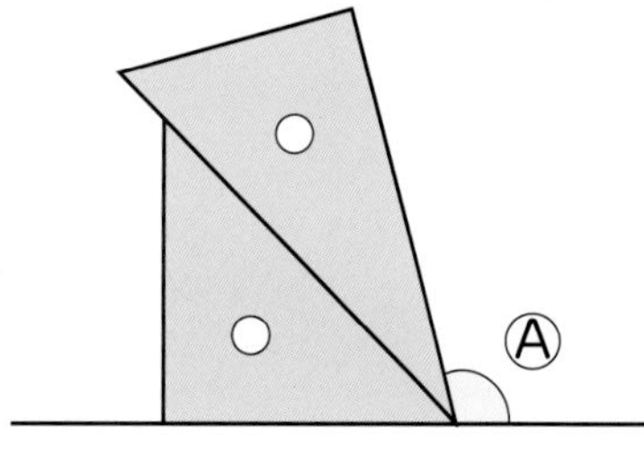

②

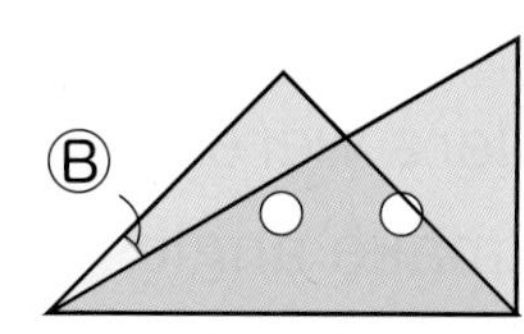

③

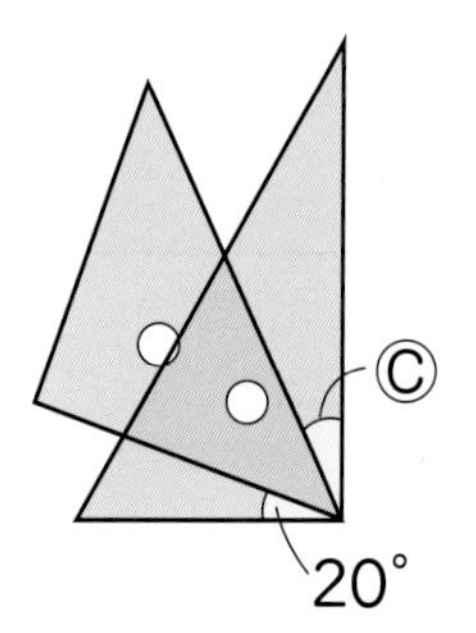

④

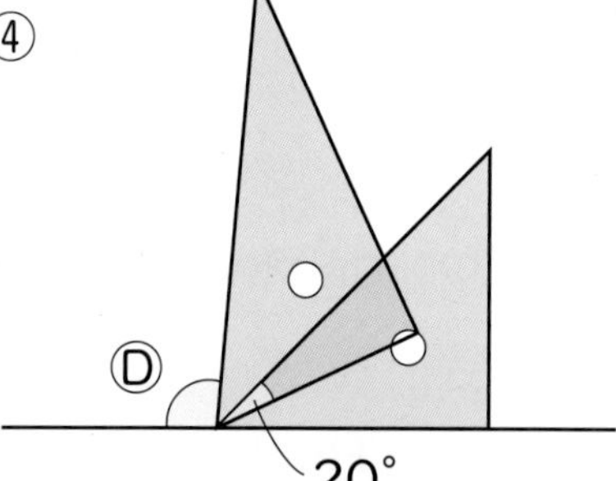

With the Way to See and Think Monsters...

Let's Reflect!

Let's reflect on which monster you used while learning "4 Angles."

Unit

We could represent the size of an angle by setting 1° as one unit, and by finding out how many of them are.

① In what way did you think when you compared the size of angles?

I set one angle of the triangle ruler as one unit. By finding out how many of them there were, I could compare the size of angles.

Haruto

If we set 1° as one unit, we could represent angles in numbers.

Akari

Divide

By dividing angles with how many more degrees than 180°, we could measure and draw large angles.

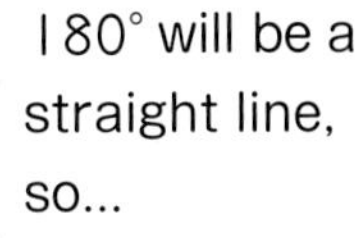

180° will be a straight line, so...

Yu

② In what way did you think when you measured the size of an angle larger than 180°?

ⓐ

Solve the ?

In the same way as for length or weight, angles could be represented in numbers by setting one unit and finding out how many units consist the size of it.

Akari

Want to Connect

What are the sizes of various triangles and quadrilaterals?

Haruto

Division with 2-digit numbers?

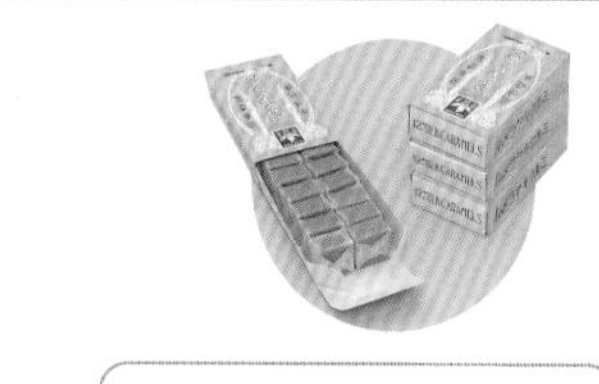

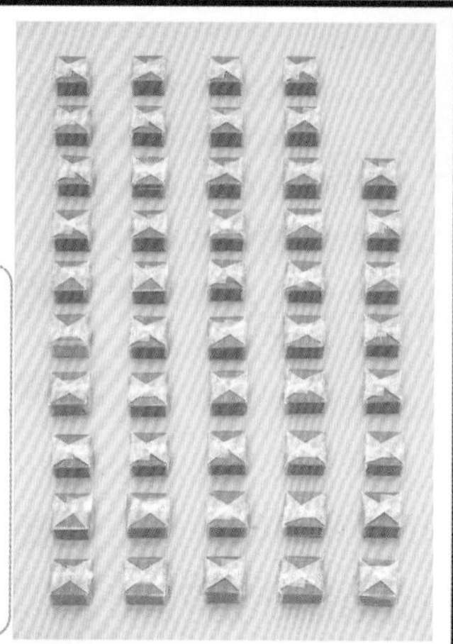

1

There are 4 boxes of caramels. Each box contains 12 caramels. We want to divide these equally.

2

To how many people?

What number is easy to divide?

3

If we divide to 2 people, how many will each person get?

We divide 4 boxes to 2 people, so, $4 \div 2 = 2$, each person will get 2 boxes. Therefore, $12 \times 2 = 24$, each person will get 24 caramels.

So we can say that $48 \div 2$ is 24.

4

If we divide among 4 people, how many will each person get?

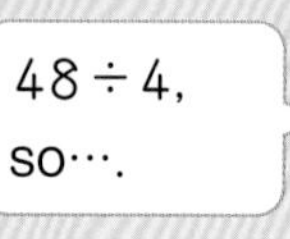

$48 \div 4$, so….

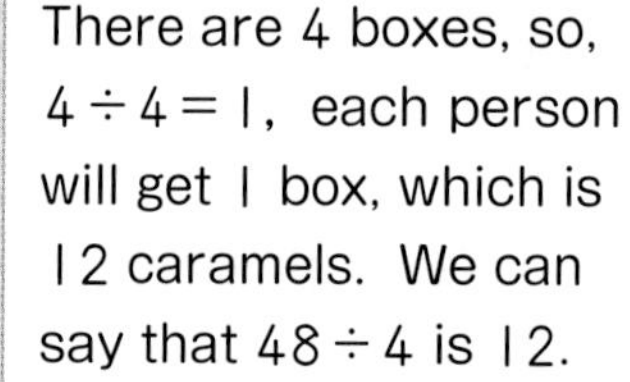

There are 4 boxes, so, $4 \div 4 = 1$, each person will get 1 box, which is 12 caramels. We can say that $48 \div 4$ is 12.

5

Actually, we are going to divide among 3 people.

$48 \div 3$, but we can't divide 4 boxes among 3 people.

How can we divide?

Want to think

How can we calculate $48 \div 3$?

The case: (2-digit) ÷ (1-digit)

Let's think about how to calculate.

(2-digit) ÷ (1-digit) →

1 There are 4 boxes of caramels. Each box contains 12 caramels. We want to divide all 48 caramels equally among 3 children. How many caramels will one child receive?

① Let's write a math expression.

☐ ÷ ☐

Total number　Number of children

Can we use the multiplication table?

Akari

② About how many caramels will one child receive?

Would the answer be a number larger than 10?

Haruto

③ Let's think about how to calculate based on what you have learned.

Let's think about various calculation methods and explain using diagrams and math expressions.

Sara's idea

First, divide 1 box each to 3 children. Next, distribute 1 box (12 caramels) to 3 children.

$12 \div 3 = 4$

One box contains 12 caramels, so,

$12 + 4 = 16$

for each child	for each child	for each child
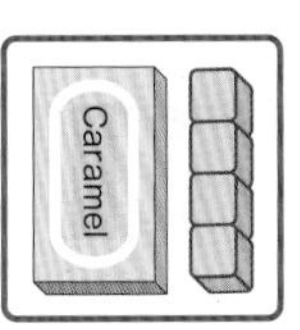	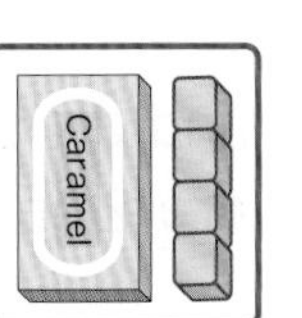	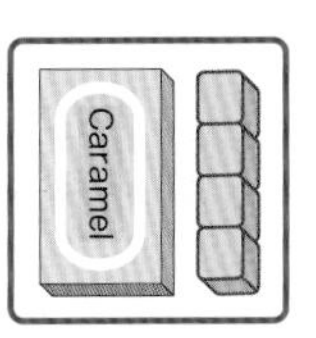

Haruto's idea

I thought about the number of caramels per box, and dividing them to 3 children. One box contains 12 caramels. Divide these with 3 children, so

$12 \div 3 = 4$

The number for each is 4.

There are 4 boxes, so, $4 \times 4 =$ ☐

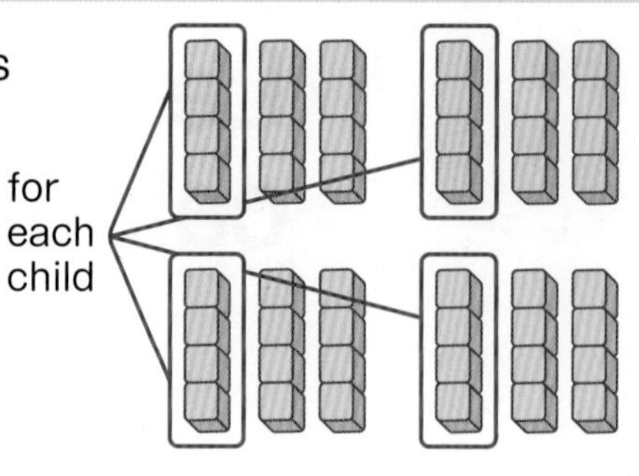

Akari's idea

I thought about the number of caramels of two boxes, and dividing them to 3 children. 2 boxes contains $12 \times 2 = 24$, so 24 caramels. Divide these with 3 children, so, $24 \div 3 = 8$. We have two more boxes, so,

$8 \times 2 =$ ☐　($8 + 8 =$ ☐)

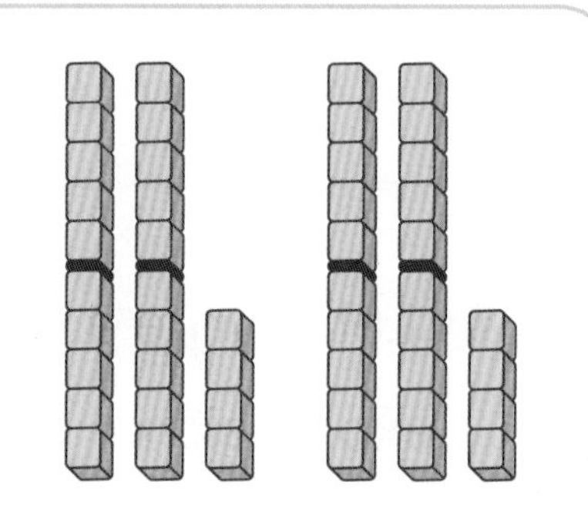

Yu's idea

I divided all the caramels into single pieces. First, divide 10 to each child, and the remainder is 18. Divide these 18 among 3 children, so,

$18 \div 3 = 6$

$10 + 6 =$ ☐

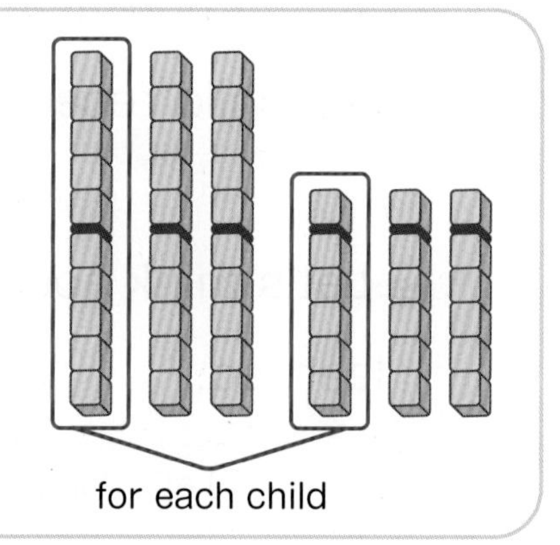

Daiki's idea

I used the multiplication table with the answer 48, $8 \times 6 = 48$.
Then, I aligned the blocks with the shape of 8×6 and divided them into 3 sets. Since there are 2 sets of 8,

$8 \times 2 =$ ☐

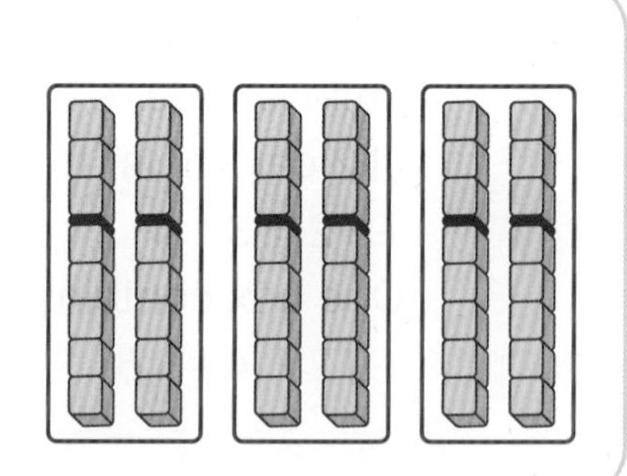

Nanami's idea

I thought about it using the rule of the divisor and the quotient.

$48 \div 3 = \square$

$\uparrow \div 2 \quad \uparrow \times 2$

$48 \div 6 = \quad 8$

When the number of dividend are the same, dividing the divisor by 2 doubles the quotient, so,

$8 \times 2 = \square$

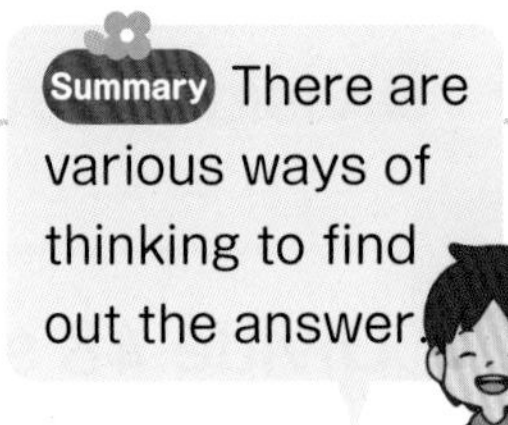

Summary There are various ways of thinking to find out the answer.
Yu

Let's think about how to calculate $56 \div 4$.

With the Way to See and Think Monsters...

Let's Reflect!

Let's reflect on which monster you used in "5 The case: (2-digit) ÷ (1-digit)."

Divide

By dividing the dividend, we could find out the answer using the multiplication table.

Rule

By using the rule of divisor and the quotient, we could find out the answer.

① How did you find out the answer for divisions such as $48 \div 3$, where the quotient is larger than 10?

I thought about dividing the number 48 into various ways.

Sara

Akari

I used the rule of division I have learned previously.

Solve the ?

By using the rules that we have learned previously and dividing the number 48, we could find out the answer for divisions where the quotient is larger than 10.

Akari

Want to Connect

Can we calculate division with larger numbers in the same way?

Haruto

Division by 1-digit Numbers

Let's think about how to calculate in vertical form.

1 Division with 1-digit quotient

1 Divide 48 sheets of colored paper into □ sheets per child. How many child will it be divided among and how many sheets will be left over?

① Let's write a math expression setting the number of sheets for each children as □.

② If you put 5, 6, 7, 8, or 9 in the □, which number is not divisible?

③ Let's set the number of sheets for each children as 9. Calculate 48 ÷ 9 in vertical form.

Division algorithm to calculate 48 ÷ 9 in vertical form ▷

Write and calculate as shown on the right.

(1) Write 5 above the ones place of 48.

(2) Write 45, from "9 times 5 is 45," in the aligned place below 48.

(3) Subtract 45 from 48. The remainder is 3.

(4) Confirm that the remainder 3 is smaller than the divisor 9.

Yu: "9 times 6 is 54" becomes larger than 48. Then, since "9 times 5 is 45," write 5.

The number of sheets of paper that can be divided is 45.

$$9\overline{)48}$$

Divide → Multiply → Subtract

$$\begin{array}{r} 5 \\ 9\overline{)48} \\ 45 \\ \hline 3 \end{array}$$

\ Want to know /

Purpose Can we calculate in vertical form with other numbers?

Haruto

④ Let's set the number of sheets for each children as 8. Calculate $48 \div 8$ in vertical form.

$8\overline{)48}$

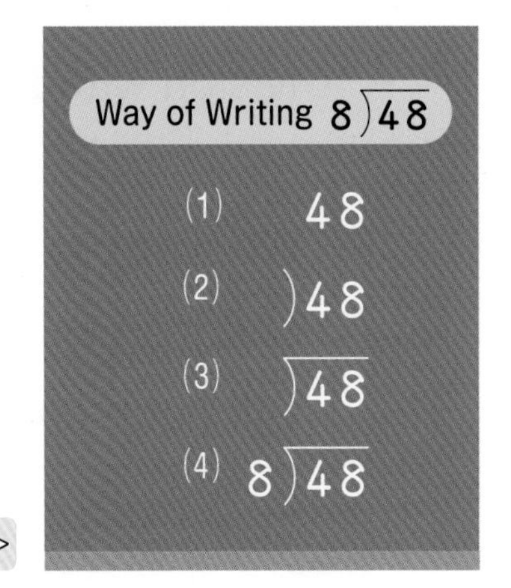

Way of Writing 8)48
(1) 48
(2))48
(3))48
(4) 8)48

In case of division with no remainder like $48 \div 8$, we can solve it in vertical form.

The answer for divisions with remainder becomes a **quotient** and a **remainder**.

1 Let's confirm the answers of the following.

① $48 \div 8 = 6$

Dividend　Divisor　Quotient

$8 \times 6 = \square$

Divisor　Quotient　Dividend

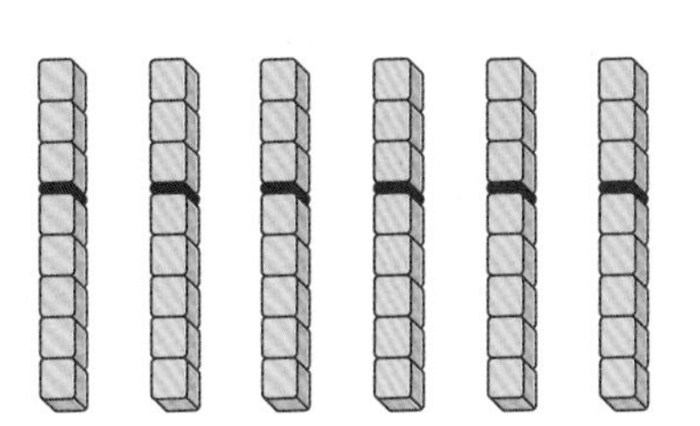

② $48 \div 9 = 5$ remainder 3

Dividend　Divisor　Quotient　Remainder

$9 \times 5 + 3 = \square$

Divisor　Quotient　Remainder　Dividend

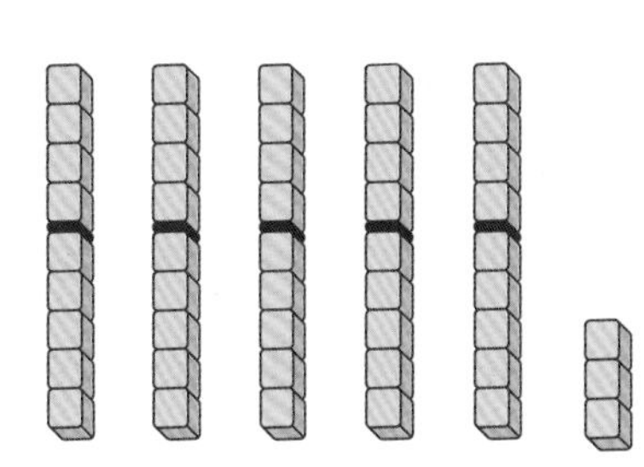

2 Let's calculate the following in vertical form. Let's confirm the answers.

① $48 \div 5$　② $48 \div 6$　③ $48 \div 7$

We can calculate division in vertical form.

Akari

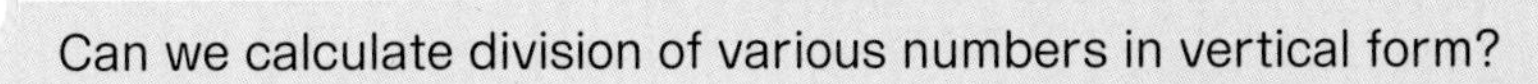
Can we calculate division of various numbers in vertical form?
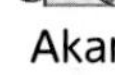

2 Division with 2-digit quotient

1 We want to divide 69 sheets of colored paper so that 3 children will receive equally. How many sheets of paper will each child receive?

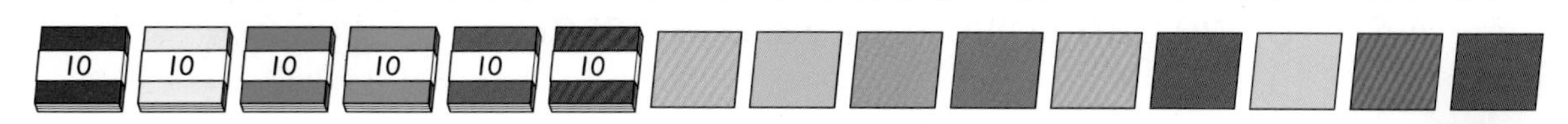

① Let's write a math expression. $\square \div \square$ (Total number of sheets ÷ Number of children)

About how many sheets?
Haruto

② Let's think about how to find the quotient of $69 \div 3$ by looking at the diagram on the right.

$$69 \div 3 \left\langle \begin{array}{l} 60 \div 3 = \square \\ 9 \div 3 = \square \end{array} \right.$$

Total $\square$

Tens	Ones
10 10	
10 10	
10 10	

▶1 We want to divide 72 sheets of colored paper so that 3 children will receive the same number. How many sheets of paper will each child receive?

10 10 10 10 10 10 10

① Let's write a math expression. $\square \div \square$ (Total number of sheets ÷ Number of children)

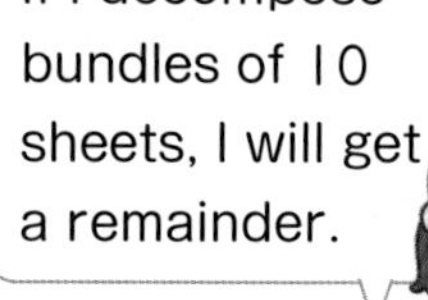

If I decompose bundles of 10 sheets, I will get a remainder.

② Let's think about how to calculate.

\ Want to know /

Purpose How should I calculate division when the tens place is not divisible?

How to think $72 \div 3$

(1) If the 7 bundles of 10 sheets are divided among 3 children, how many bundles will each child receive and how many bundles will remain?

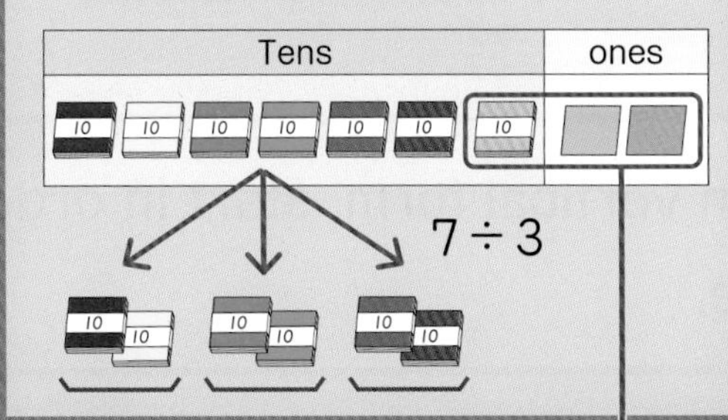

(2) Break up the remaining 1 bundle and combine that with the 2 single sheets.

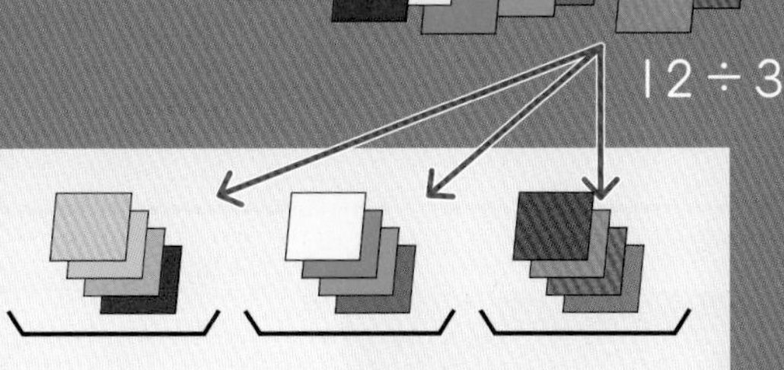

(3) Divide the 12 sheets among 3 children.

(4) As for each child,

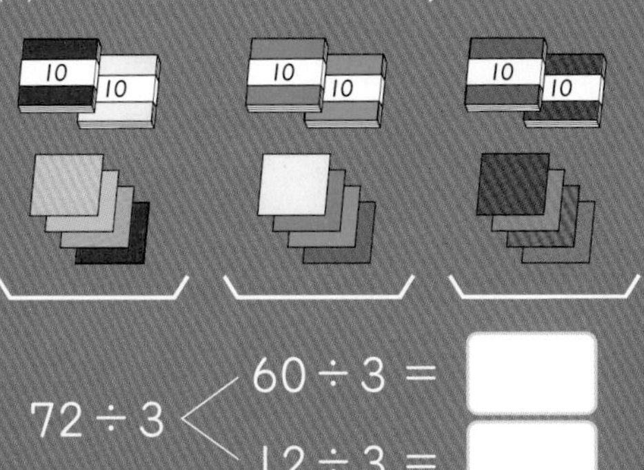

$72 \div 3$ < $60 \div 3 = \square$, $12 \div 3 = \square$

Total $\square$

Division algorithm to calculate $72 \div 3$ in vertical form

Tens place calculation

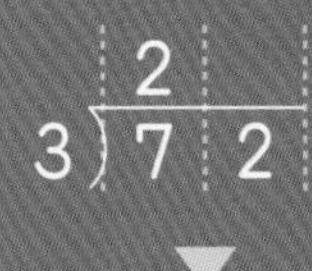

Divide: $7 \div 3 = 2$ remainder 1. Write 2 in the tens place.

Multiply: $3 \times 2 = 6$

The 6 means that we are using 6 bundles of 10 sheets.

Subtract: $7 - 6 = 1$

The remainder is smaller than the divisor.

Bring down: Bring down the 2 in the ones place.

Ones place calculation

Divide: $12 \div 3 = 4$. Write 4 in the ones place.

Multiply: $3 \times 4 = 12$

The 12 means that we are using 12 single sheets.

Subtract: $12 - 12 = 0$

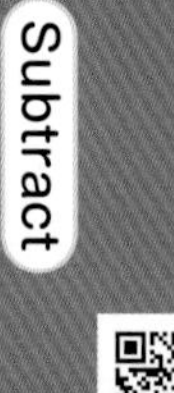

③ Let's calculate $72 \div 3$ in vertical form.

Try it by yourself by looking at the vertical form in the previous page.

Summary

When calculating divisions in vertical form, start in order from the highest place value.

? When we think about the number to write on the tens place, what do we pay attention to?

Explaining division with 2-digit quotient ↓

2

Sakura was solving $92 \div 4$ in vertical form. In the process she was in trouble. Let's think about why she was in trouble, and calculate correctly.

＼ Want to tell ／

Purpose Let' s explain to your friends.

Haruto

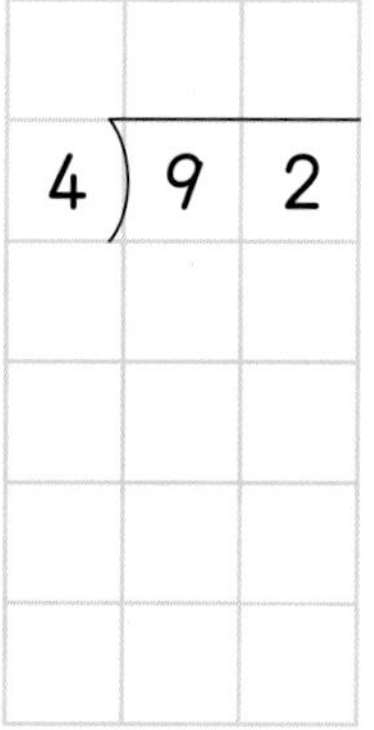

1 Let's calculate the following in vertical form.

① $54 \div 2$　② $68 \div 4$　③ $84 \div 3$　④ $74 \div 2$

? Is there a case that is not divisible in a division where the quotient becomes 2-digit number?

3

We want to divide 83 sheets of colored paper so that 5 children will receive the sheets equally. How many sheets of paper will each child receive and how many sheets of paper will remain?

① Let's write a math expression. ☐

② Let's explain the division algorithm in vertical form.

\ Want to think /

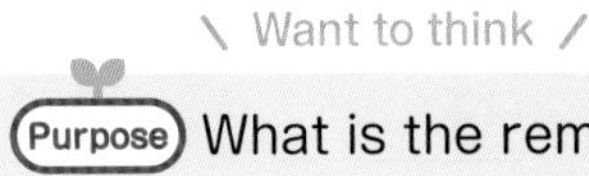

Purpose What is the remainder?

Let's confirm that the remainder becomes smaller than the divisor.

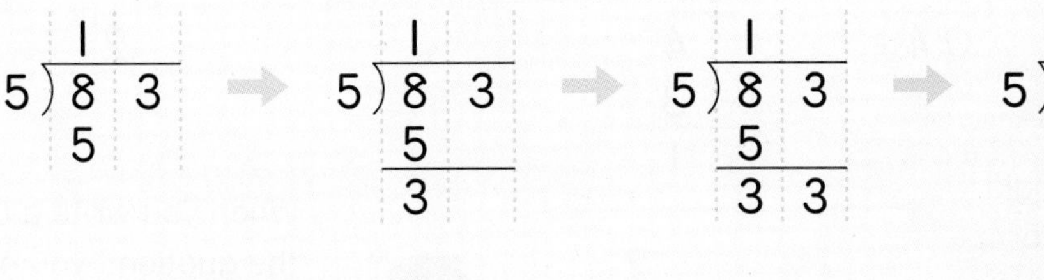

$8 \div 5 = 1$ remainder 3
Write 1 in the tens place. $5 \times 1 = 5$

$8 - 5 = 3$

Bring down the 3 in the ones place.

$33 \div 5 = 6$ remainder 3
Write 6 in the ones place.
$5 \times 6 = 30$ $33 - 30 = 3$

$83 \div 5 =$ ☐ remainder ☐

Answer: Each child receives ☐ sheets and the remainder is ☐ sheets.

③ Let's confirm the answer.

5 × ☐ + ☐ = ☐

Divisor × Quotient + Remainder = Dividend

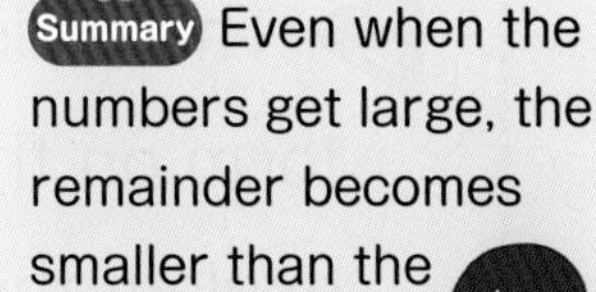

Summary Even when the numbers get large, the remainder becomes smaller than the divisor.
Akari

1 Let's calculate the following in vertical form. Let's confirm the answers.

① $37 \div 2$ ② $58 \div 4$ ③ $86 \div 7$

④ $93 \div 8$ ⑤ $76 \div 6$ ⑥ $47 \div 3$

2 There is a tape that is 67 cm long. How many 5 cm long pieces can you make? How many cm will the remainder be?

? What should I do if the tens place is divisible?

4 Let's explain the following division algorithm in vertical form.

❶ $64 \div 3$

```
   2          2          21
3)64  →   3)64  →   3)64
             6          6
            04          4
                        3
                        1
```

❷ $92 \div 3$

```
   3          30          30
3)92  →   3)92  →   3)92
             9           9
             2           2
                         0
                         2
```

\ Want to know /

Purpose How should I do in vertical form if the tens place is divisible?

Sara

You don't have to write the 0.

```
   2
3)64
  6
  04
```

Yu

When you write a 0 in the quotient, you can omit the calculation inside.

```
   30
3)92
  9
   2
```

1 Let's calculate the following in vertical form.

① $45 \div 2$　② $97 \div 3$　③ $57 \div 5$

④ $62 \div 3$　⑤ $81 \div 2$　⑥ $82 \div 4$

2 The division algorithms in vertical form on the right are wrong. Let's explain the reasons and solve them correctly.

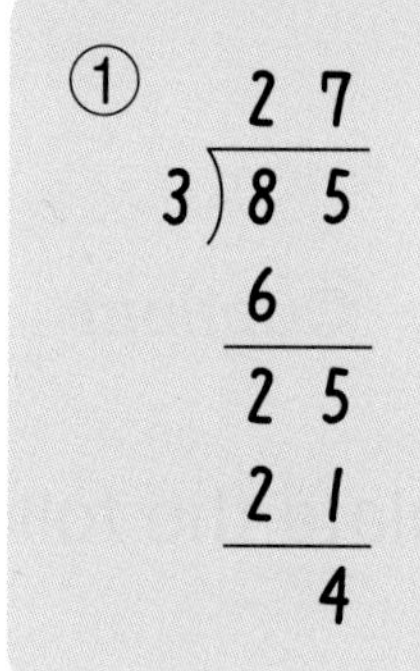

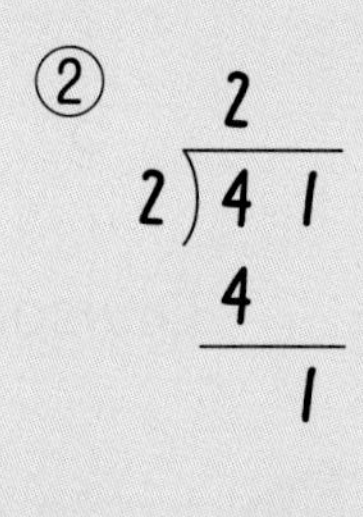

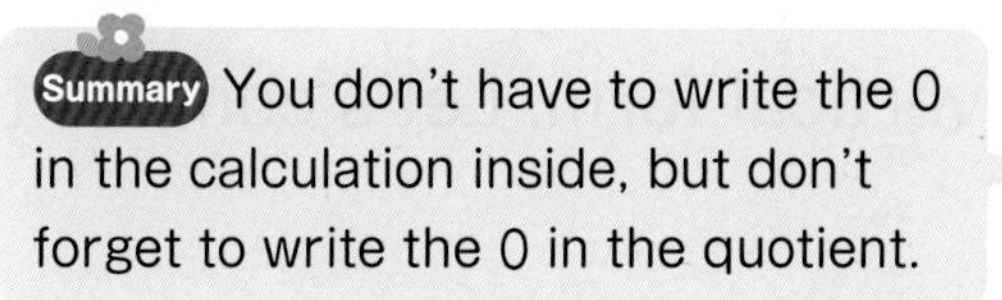

Summary You don't have to write the 0 in the calculation inside, but don't forget to write the 0 in the quotient.

Akari

? Can we calculate division in vertical form with large numbers?

3 The case: (3-digit) ÷ (1-digit)

1 There are 639 sheets of colored paper. If the papers are divided so that 3 groups will receive the sheets equally, how many sheets of paper will each group receive?

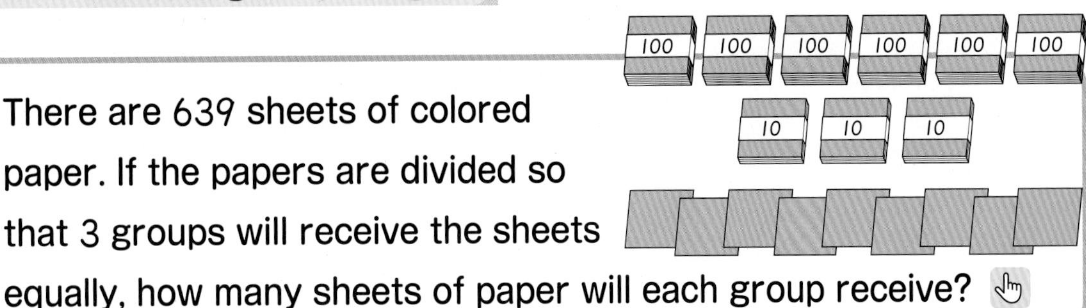

① Let's write a math expression.

$639 \div 3$ — $600 \div 3 = \square$, $30 \div 3 = \square$, $9 \div 3 = \square$; Total $\square$

② About how many hundred sheets of paper would it each group get?

③ Let's think about the division algorithm in vertical form.

1▶ We want to divide 536 sheets of colored paper so that 4 children will receive the sheets equally. How many sheets of paper will each child receive? Let's think about how to calculate. $536 \div 4$

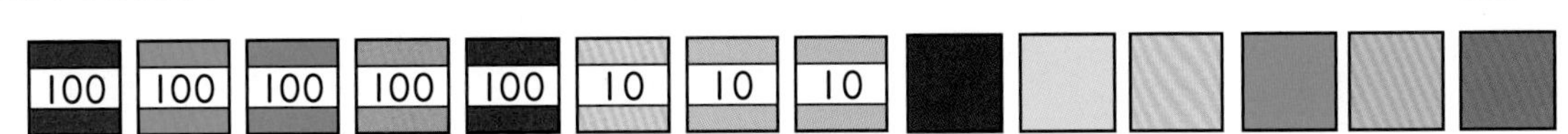

① Let's divide the bundles of 100 sheets.

$5 \div 4 = \square$ remainder $\square$

Number of bundles of 100 sheets

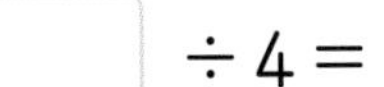

If we set the remaining 100-sheet bundles into 10-sheet bundles, how many 10-sheet bundles will I get?

Akari

② Let's divide the bundles of 10 sheets. $\square \div 4 = \square$ remainder $\square$

③ Let's divide the single sheets. $\square \div 4 = \square$

④ How many sheets of paper will each child receive? $536 \div 4 = \square$

⑤ Let's think about the division algorithm in vertical form.

Haruto

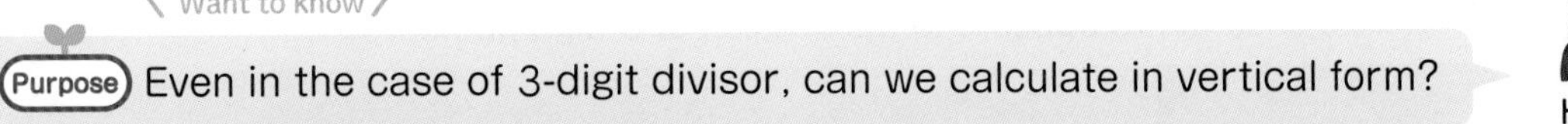

\ Want to know /

Purpose Even in the case of 3-digit divisor, can we calculate in vertical form?

Division algorithm to calculate 536 ÷ 4 in vertical form

From which place did we divide?

Hundreds	Tens	Ones
100 100 100 100 (100 →)	100 10 10 10	

5 ÷ 4

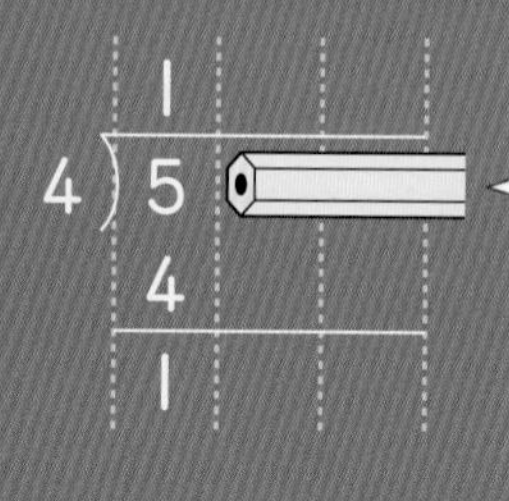

Divide the number of bundles of 100 sheets. $4\overline{)5}$

Hundreds	Tens	Ones
	10 10 10 10 10 10 10 10 10 10 10 10 (10 →) 10	

13 ÷ 4

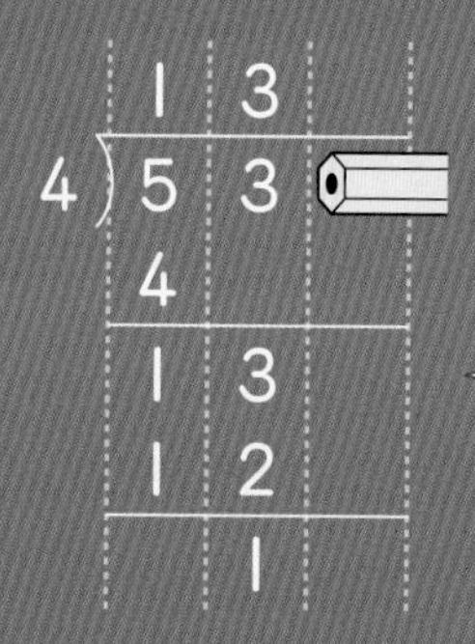

Divide the number of bundles of 10 sheets. $4\overline{)13}$

Hundreds	Tens	Ones

16 ÷ 4

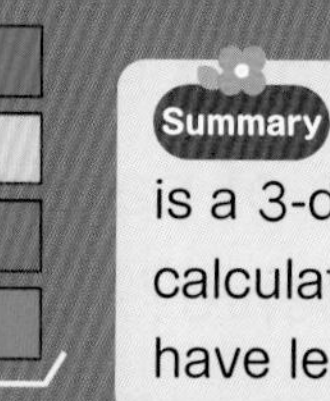

Divide the number of single sheets. $4\overline{)16}$

Summary Even when the dividend is a 3-digit number, we can calculate in vertical form as we have learned before.

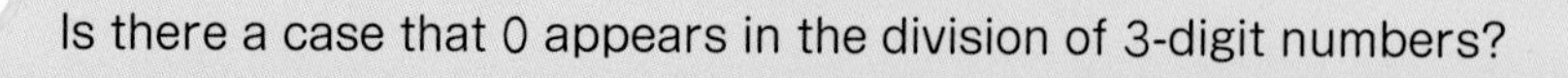

Is there a case that 0 appears in the division of 3-digit numbers?

2 Let's explain the following divisions Ⓐ and Ⓑ in vertical forms. Let's confirm the answers.

\ Want to know /

Purpose What should I do if it was divisible during the process?

1 $420 \div 3$

Ⓐ
```
    1 4 0
3 ) 4 2 0
    3
    1 2
    1 2
        0
        0
        0
```

Ⓑ
```
    1 4 0
3 ) 4 2 0
    3
    1 2
    1 2
        0
```

2 $859 \div 8$

Ⓐ
```
    1 0 7
8 ) 8 5 9
    8
      5
      0
      5 9
      5 6
        3
```

Ⓑ
```
    1 0 7
8 ) 8 5 9
    8
      5 9
      5 6
        3
```

1 Let's calculate the following in vertical form.

① $740 \div 2$　② $650 \div 5$

③ $742 \div 7$　④ $958 \div 9$

2 Let's explain the mistakes in the division algorithm in vertical form shown on the right. Calculate correctly.

```
      2 7
3 ) 6 2 2
    6
      2 2
      2 1
        1
```

Summary Pay attention on how to treat 0 like we did in 2-digit numbers.
Sara

? In a division to divide a 3-digit number, is there a case that the answer becomes a 2-digit number?

That's it! **Mental Calculation** Let's calculate $72 \div 4$ mentally.

$72 \div 4$ <
- $40 \div 4$ → "4 and 1 is 4" → 10
- $32 \div 4$ → "4 and 8 is 32" → 8

> → Total ☐

4 The case: (3-digit) ÷ (1-digit) = (2-digit)

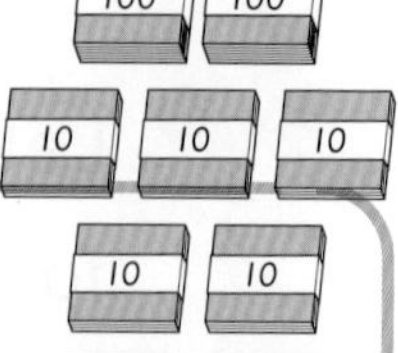

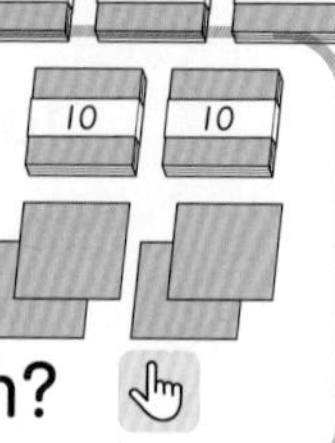

1 We want to divide 254 sheets of colored paper so that 3 children will receive equally. How many sheets will each child receive and how many sheets will remain?

① Let's write a math expression.

② Let's think about the division algorithm in vertical form.

\ Want to know /

Purpose What should we do in a calculation when the quotient cannot be written in the hundreds place?

Division algorithm to calculate 254 ÷ 3 in vertical form

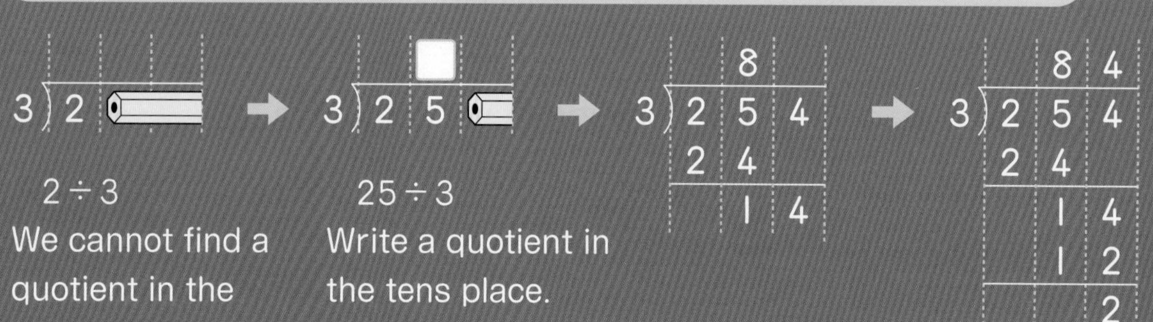

2 ÷ 3
We cannot find a quotient in the hundreds place.

25 ÷ 3
Write a quotient in the tens place.

Summary When the quotient is not written in the hundreds place, the calculation starts from the tens place.

③ Let's confirm the answer.

1 Let's calculate the following in vertical form.

① 316 ÷ 4　② 552 ÷ 6　③ 329 ÷ 7　④ 624 ÷ 8

⑤ 173 ÷ 2　⑥ 581 ÷ 9　⑦ 236 ÷ 3　⑧ 488 ÷ 5

CAN What can you do?

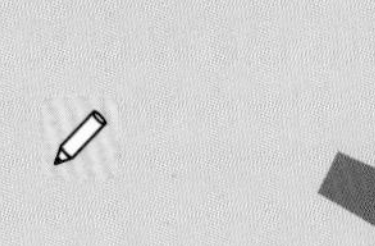

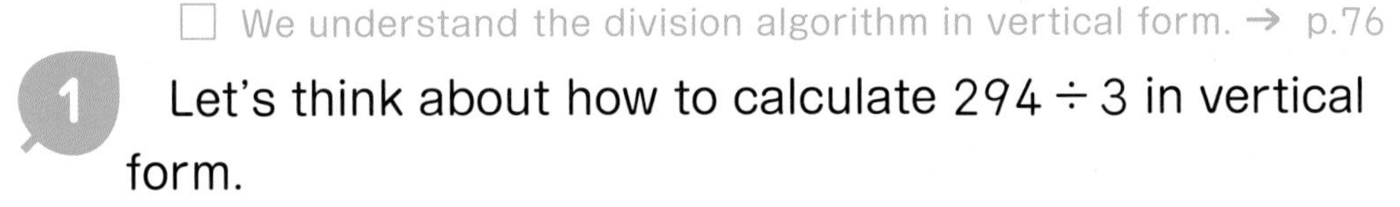

☐ We understand the division algorithm in vertical form. → p.76

1 Let's think about how to calculate 294 ÷ 3 in vertical form.

$3\overline{)2\ 9\ 4}$

① The quotient is written from the ☐ place value.

② The remainder 2 in the tens place represents 2 sets of ☐ .

③ The calculation in the ones place is ☐ ÷ 3.

☐ We can calculate division in vertical form. → pp.68 ~ 76

2 Let's calculate the following in vertical form.

① 78 ÷ 3	② 96 ÷ 8	③ 38 ÷ 2	④ 55 ÷ 5
⑤ 48 ÷ 4	⑥ 77 ÷ 6	⑦ 56 ÷ 3	⑧ 90 ÷ 7
⑨ 83 ÷ 2	⑩ 65 ÷ 3	⑪ 98 ÷ 9	⑫ 81 ÷ 4
⑬ 548 ÷ 4	⑭ 259 ÷ 7	⑮ 624 ÷ 3	⑯ 367 ÷ 9
⑰ 457 ÷ 6	⑱ 543 ÷ 5	⑲ 963 ÷ 8	⑳ 728 ÷ 6

☐ We can make a division expression and find out the answer. → pp.68 ~ 76

3 Let's answer the following problems.

① Mitsuki and her 5 friends will fold 372 cranes. If each child folds the same number of cranes, how many cranes should each one fold?

② There is a 64cm rope. We want to make the shape of a square. How long will the length of each sides be in cm?

Supplementary Problems → p.146

Which "Way to See and Think Monsters" did you find in "6 Division by 1-digit Numbers"?

I found "Divide" when I was trying to calculate division in large numbers.

Sara

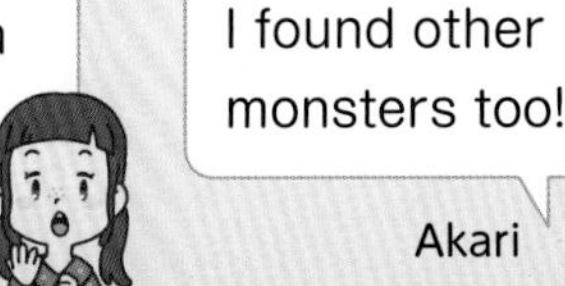

I found other monsters too!

Akari

With the Way to See and Think Monsters...

Let's Reflect!

Let's reflect on which monster you used in "6 Division by 1-digit Numbers."

Divide

Even when the numbers get large, by dividing into each place, we could calculate using the multiplication table.

① How did you calculate the following?

$69 \div 3$ → $60 \div 3 = \square$, $9 \div 3 = \square$, Total $\square$

$639 \div 3$ → $600 \div 3 = \square$, $30 \div 3 = \square$, $9 \div 3 = \square$, Total $\square$

Akari

Even when the dividend is a 2-digit or 3-digit number, we divided them into each place.

② How did you calculate the division on the right?

How did you divide 917?

Haruto

```
    2 2 9
4 ) 9 1 7
    8
    1 1
      8
      3 7
      3 6
        1
```

③ What does the number "11" mean in the vertical form on the right?

Let's deepen. → p.151

Solve the ?

Even when the dividend is a 2-digit or 3-digit number, we could calculate in vertical form by dividing them into each place.

Haruto

Want to Connect

Even when the numbers get larger, can we calculate division?

Yu

Find the ?

Which place and which injury happens the most?

Want to know

What should we do to summarize and make data easy to understand?

Arrangement of Data

Let's think about how to summarize tables.

1 Arrangement of table

Arrangement of table →

1 Let's organize the data in the table on the right and explore the situation of injuries in the school.

① Let's explore the places that injuries happened.

Ⓐ Where did the injuries happened most frequently? Let's write down on the table below.

Ⓑ Look at the table and explain what you have noticed.

Number of children and place of injury

Place of injury	Number of children	
Playground	一	
Corridor	T	
Classroom		
Gymnasium		
Stairs		
Total		

Write the character "正" to organize as you did in the 3rd grade.

Record of injured children

Grade	Time	Place	Kind of injury
5	Morning	Corridor	Bruise
4	Break	Playground	Scratch
5	Break	Corridor	Bruise
1	Lunch break	Classroom	Scratch
3	During class	Playground	Scratch
3	During class	Playground	Bruise
6	During class	Gymnasium	Scratch
5	Morning	Classroom	Cut
4	Break	Playground	Scratch
5	During class	Gymnasium	Scratch
3	During class	Gymnasium	Bruise
1	During class	Classroom	Cut
2	During class	Playground	Scratch
6	During class	Gymnasium	Sprain
6	After school	Playground	Sprained finger
5	Morning	Classroom	Cut
5	Break	Classroom	Scratch
3	Break	Stairs	Bruise
4	During class	Gymnasium	Sprain
2	During class	Playground	Bruise
6	During class	Gymnasium	Scratch
4	Lunch break	Corridor	Bruise
4	Morning	Stairs	Scratch
2	During class	Gymnasium	Sprained finger
1	During class	Playground	Sprain
6	Lunch break	Corridor	Cut
5	Break	Classroom	Scratch
3	During class	Playground	Bruise

❷ Let's explore the type of injury.

Ⓐ Let's write down on the table which injuries happened most frequently.

Ⓑ Look at the table and explain what you have noticed.

Number of children and kind of injury

Kind of injury	Number of children
Cut	
Bruise	
Scratch	
Sprained finger	
Sprain	
Total	

Summary Summarizing depending on different perspectives makes data easier to understand.

Sara

? Can we put the two perspectives together and make it easier to understand?

2

Let's explore the kind of injury and where it happened. Keeping an eye on both the kind as well as the place, let's fill in the table with appropriate numbers.

Places and kinds of injuries (children)

Place \ Kind	Cut	Bruise	Scratch	Sprained finger	Sprain	Total
Playground						
Corridor						
Classroom						
Gymnasium						
Stairs						
Total						

❶ Where and which kind of injury has the largest number?

❷ Where did the most frequent kind of injury happen?

❸ Compared to the two tables summarized in 1, what did you find out in this table?

Yu

Is there a difference between the kind and place of injury depending on the grade or time zone?

What kind of poster should I make, and where should I post?

Akari

I want to have goldfish or small birds.

Why don't you ask your classmates which they have, and think about which one to choose?

2 Arrangement of data

1 Haruto asked his classmates to put a ○ on the card to investigate whether they have goldfish (G) or small birds (SB) at home. Let's answer the following questions.

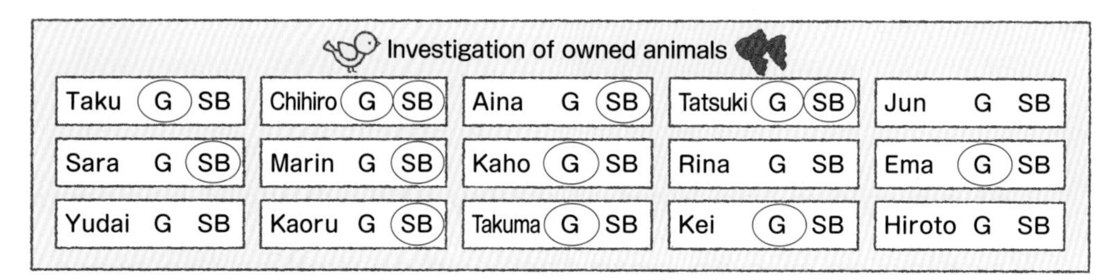

Investigation of owned animals
Taku G SB
Chihiro G SB
Aina G SB
Tatsuki G SB
Jun G SB
Sara G SB
Marin G SB
Kaho G SB
Rina G SB
Ema G SB
Yudai G SB
Kaoru G SB
Takuma G SB
Kei G SB
Hiroto G SB

① What kind of groups can we made depending on the way the ○ are marked?

Ⓐ Who drew two ○ and how many children are there in this group?

Ⓑ Who drew one ○ and how many children are there in this group?

Ⓒ Divide the children who drew one ○ into those who have goldfish and those who have small birds. How many children are there in each group?

Ⓓ Who did not put any ○ and how many children are there in this group?

② Let's fill in the following table with the corresponding number of children.

Investigation of owned animals (children)

		Goldfish		Total
		Have	Don't have	
Small birds	Have	2		
	Don't have			
Total				

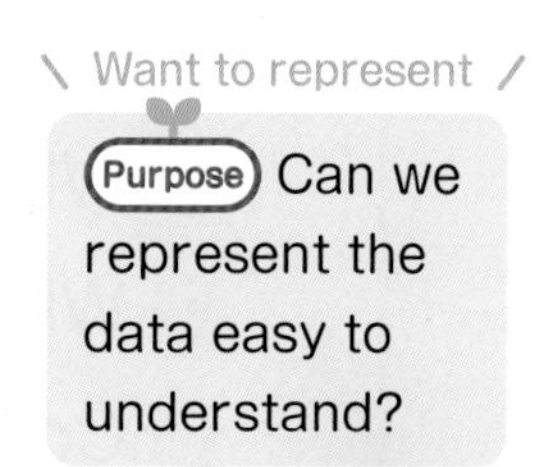

Want to represent
Purpose Can we represent the data easy to understand?

③ How many children only have birds?

④ How many children have goldfish?

CAN What can you do? 

☐ We can make the summary to understand two things. → p.81

1 The following table shows the record of injuries for the 4th grade in Soma's school. Let's summarize the data in the table below.

Record of injured children

Name	Place	Kind of injury
Mei	Playground	Scratch
Hinata	Classroom	Cut
Daiki	Classroom	Scratch
Yui	Gymnasium	Sprain
Sora	Corridor	Bruise
Mio	Gymnasium	Sprained finger

Name	Place	Kind of injury
Hiroto	Playground	Bruise
Hana	Playground	Cut
Akito	Gymnasium	Scratch
Sakura	Gymnasium	Bruise
Nanami	Classroom	Scratch
Soma	Gymnasium	Scratch

Places and kinds of injuries (children)

place \ kind	Scratch					Total
Playground						
Total						

☐ We can read a summary from a table. → p.82

2 Mitsuki investigated about the siblings of her classmates.
There are 36 children in her class.
Children with older brothers: 12 children
Children with older sisters: 9 children
Children without older siblings: 18 children
Complete the table on the right with numbers of children.

(children)

		Older brother		Total
		Yes	No	
Older sister	Yes			
	No			
Total				36

Supplementary Problems → p.147

Which "Way to See and Think Monsters" did you find in "7 Arrangement of Data"?

I found "Summarize" when I was trying to make a table with two perspectives.
Haruto

I found other monsters too!
Akari

With the Way to See and Think Monsters...

Let's Reflect!

Let's reflect on which monster you used while learning "7 Arrangement of Data".

Summarize

By summarizing the two perspectives such as kind and place, there were cases where the table turned easier to understand.

① What do you find out from the following table?

Places and kinds of injuries (children)

Place \ Kind	Cut	Bruise	Scratch	Sprained finger	Sprain	Total
Playground	0	3	4	1	1	9
Corridor	1	3	0	0	0	4
Classroom	3	0	3	0	0	6
Gymnasium	0	1	3	1	2	7
Stairs	0	1	1	0	0	2
Total	4	8	11	2	3	28

We can find out which kind of injury is the most and the least by seeing the graph horizontally.

Akari

We can find out where they tend to get injured by seeing the graph vertically.

Yu

We can find out where and what kind of injury happens by seeing both.

Haruto

Let's deepen. → p.152

? Solve the ?

By putting the two perspectives together, tables got easier to understand.

Sara

Want to connect

By using tables, bar graphs, and line graphs, can we find out various things?

Yu

Compared to the tables learned in the 3rd grade, the tables learned in the 4th grade can summarize two points of view, but it is a little more difficult to read. It will be useful to be able to read and summarize such tables correctly when organizing the data.

❶ The table on the right summarizes the results of the survey on whether or not you have a dog or cat at home. Let's answer the following questions.

① What does "1" in the table represent? Explain using the words in the table.

Animals that children have (children)

		Cat		Total
		○	×	
Dog	○	2	6	Ⓑ
	×	1	4	Ⓒ
Total		Ⓓ	Ⓔ	Ⓐ

○…have ×…do not have

The box above "1" represents ○ for cats, and the box on the left of "1" represents × for dogs...

Akari

Frequently made mistake
Explaining only one or the other, as in "the number of children who have cats" or "the number of children who don't have dogs."

Be careful!
The "1" in the table represents the number of children who have cats and do not have dogs. Be careful how you read the table.

		Cat	
		○	×
Dog	○	2	6
	×	1	4

② Let's find out the number that fits in Ⓐ.

Haruto

Since "total" and "total" have just been combined, should I just count all the numbers in the "total" part?

		Cat		Total
		○	×	
Dog	○	2	6	Ⓑ8
	×	1	4	Ⓒ5
Total		Ⓓ3	Ⓔ10	Ⓐ

8+5+3+10 ?

Frequently made mistake
Write the sum of Ⓑ, Ⓒ, Ⓓ, and Ⓔ in Ⓐ.

Be careful!
Ⓐ represents the total number of children who responded to the investigation. The sum of Ⓑ and Ⓒ and the sum of Ⓓ and Ⓔ will be the same, and that number goes into Ⓐ.

When reading the table summarizing the two perspectives, pay attention to what each number represents.

Reflect

Connect

Problem In the target game, what is the tendency of getting on the target in the 1st and 2nd turn?

Name	Tatsuya	Mamoru	Aiko	Yuuki	Takeshi	Kana	Atsushi	Kokoro
1st turn	◎	○	◎	◎	○	△	○	◎
2nd turn	◎	◎	◎	◎	△	○	○	○
Name	Yuki	Shizuka	Kazue	Rika	Yuko	Maya	Ken	Yui
1st turn	◎	◎	△	△	○	◎	◎	○
2nd turn	○	◎	△	△	○	◎	◎	◎
Name	Keiko	Manabu	Airi	Wakana	Daisuke	Nami	Yoshiharu	Takahiro
1st turn	○	○	◎	△	○	○	◎	△
2nd turn	○	◎	◎	○	◎	△	◎	◎
Name	Yuma	Sota	Kazuhiro	Hiroki	Noriko	Toshihiro		
1st turn	◎	○	△	◎	○	◎		
2nd turn	○	○	○	○	◎	◎		

It is hard to see the numbers of ◎, ○, and △ for each turn.

As for the first table, I can understand the person and the tendency of getting on the target, but it is not easy to see the numbers of ◎, ○, and △.

Is there another way to make the table easier to see?

[Target]

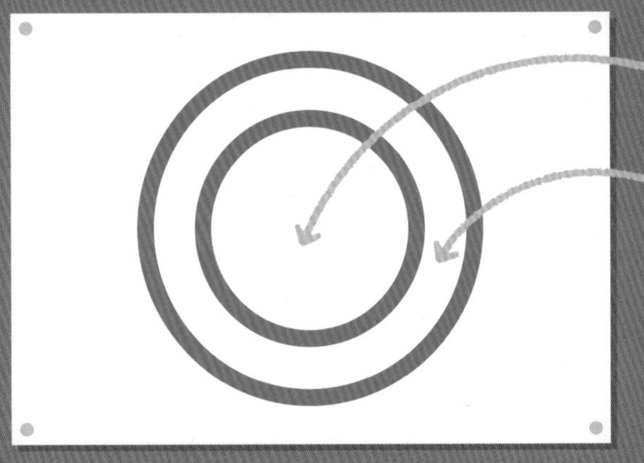

If you hit here: ◎

If you hit here: ○

If you do not hit: △

It is easy to understand because the 1st and 2nd turn are placed vertically and horizontally respectively.

		2nd turn			Total
		◎	○	△	
1st turn	◎	9	4	0	13
	○	5	4	2	11
	△	1	3	2	6
Total		15	11	4	30

Summary

- The largest number of children hit ◎ for both 1st and 2nd turn.
- The second largest number of children hit ○ for the 1st turn and ◎ for the 2nd turn.
- There are no children who got ◎ for the 1st turn and △ for the 2nd turn.

Can we make a table like this which I have made before?

		Goldfish		Total
		Have	Don't have	
Small birds	Have			
	Don't have			
Total				

Want to Connect

As the number of items increases, can we make a table that is easy to see by adding columns and rows?

Division by 2-digit numbers?

Let's divide these 80 sheets of colored paper equally.
10 10 10 10 10 10 10 10
How many each?
To how many students?
1

If we distribute 5 sheets of paper, it will be 80 ÷ 5.

We can calculate in vertical form.

$$5\overline{)80}$$

2

I want to distribute 10 or 20 or 30 sheets to each student.

If we want to distribute 10, we can divide one bundles each.

If we want to distribute 20, it will be 80 ÷ 20 ?

If we want to distribute 30, it will be 80 ÷ 30?

Can we think about making bundles of 10 sheets?

3

How can you calculate a division by a 2-digit numbers?

Division by 2-digit Numbers

Let's think about how to calculate in vertical form.

1 Division by tens

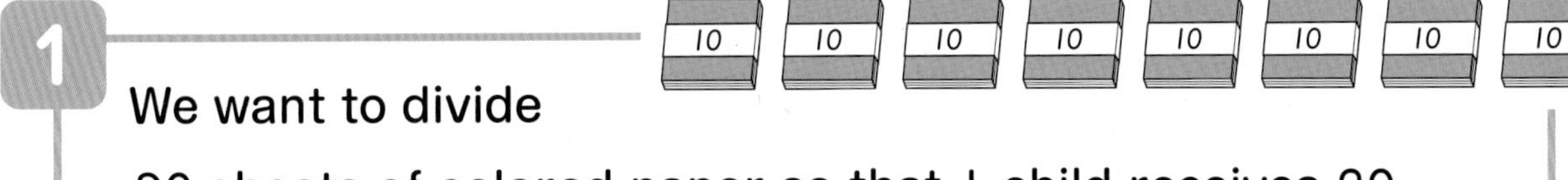

1 We want to divide 80 sheets of colored paper so that 1 child receives 20 sheets of paper. How many children can share the sheets?

1. Let's write a math expression. ☐
2. Let's think about how to calculate.

\ Want to know /

Purpose How can we calculate division by tens?

Sara's idea

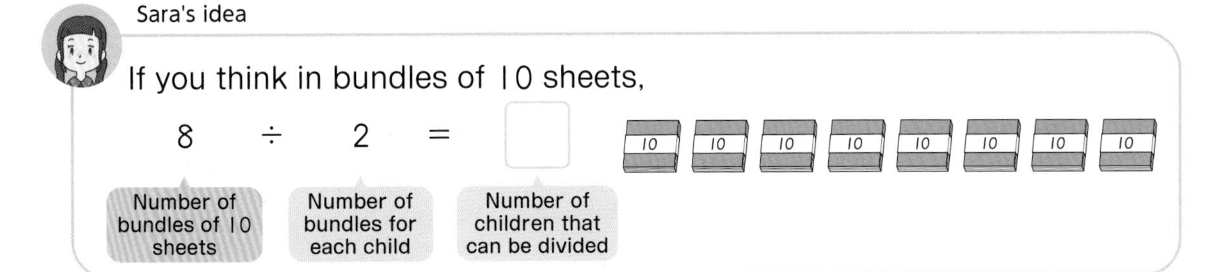

If you think in bundles of 10 sheets,
8 ÷ 2 = ☐
Number of bundles of 10 sheets
Number of bundles for each child
Number of children that can be divided

80 ÷ 20 = ☐　　Answer: ☐ children

Way to see and think

Summary

As for 80÷20, if you think based on 10, you can find the answer by calculating 8 ÷ 2.

1 Let's calculate the following.

① 40 ÷ 20　② 60 ÷ 30　③ 90 ÷ 30　④ 80 ÷ 10

⑤ 150 ÷ 50　⑥ 120 ÷ 20　⑦ 240 ÷ 60　⑧ 160 ÷ 80

? In case of 80 ÷ 30, should we think by bundles of 10?

2 We want to divide 80 sheets of colored paper so that 1 child receives 30 sheets of paper. How many children can share the sheets?

① Let's write a math expression. [ ]

② Let's think about how to calculate.

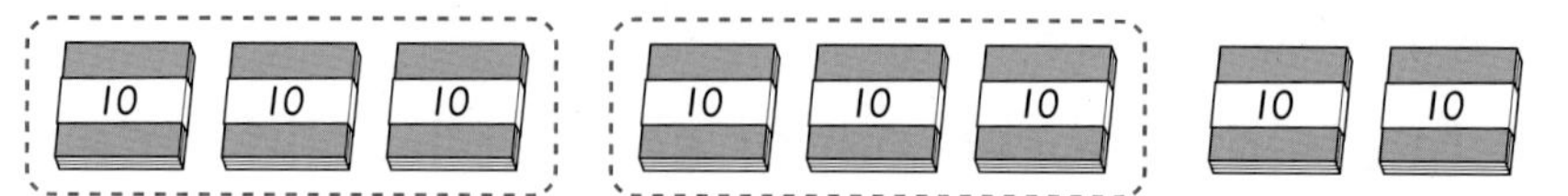

If we think in bundles of 10, $8 \div 3 = 2$ remainder 2. — Haruto

Is the remainder 2?
Akari

The remainder is 2 bundles of 10, therefore... — Yu

\ Want to think /

Purpose If we think of bundles of 10, what will the remainder be?
Sara

Answer: [ ] children can receive and the remainder is [ ] sheets

③ Let's confirm the answer.

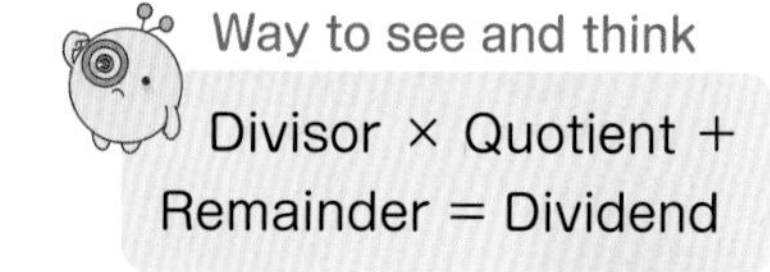

Way to see and think
Divisor × Quotient + Remainder = Dividend

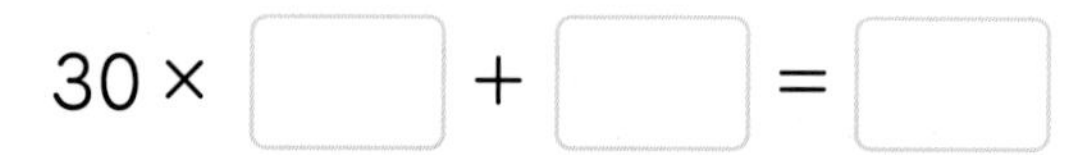

Summary They are calculating in bundles of 10, so we need to think about the remainder in bundles of 10.

1 We want to divide 140 sheets of colored paper so that 1 child receives 30 sheets of paper. How many children can share the sheets? How many sheets will remain?

If we think of bundles of 10, it will be $14 \div 3$, so...

2 Let's calculate the following.

① $70 \div 20$	② $90 \div 40$	③ $50 \div 20$	④ $60 \div 50$
⑤ $320 \div 60$	⑥ $180 \div 70$	⑦ $200 \div 30$	⑧ $250 \div 80$

? Can we do division when both the dividend and the divisor is not a number wth bundles of 10?

The case: (2-digit) ÷ (2-digit) ↓

2 Division by 2-digit number (1)

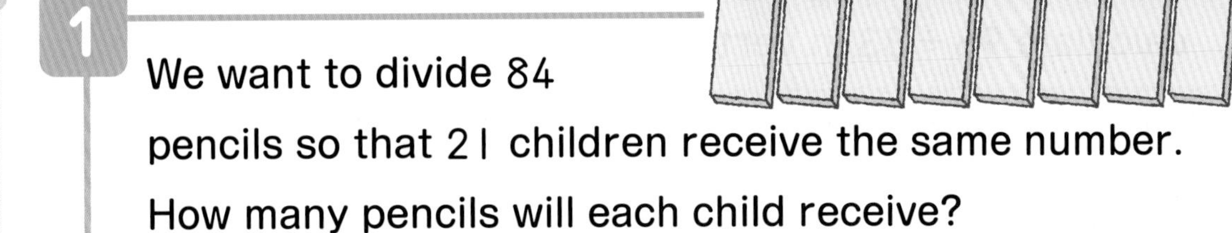

1 We want to divide 84 pencils so that 21 children receive the same number. How many pencils will each child receive?

① Let's write a math expression. ☐

② If you calculate in vertical form, from which place value is the quotient written?

$21\overline{)84}$ ☐

$2\square\overline{)8\square}$ 4

\ Want to think /

Purpose Can we calculate division by 2-digit number in vertical form?

Akari

I can't do $21\overline{)8}$.

Haruto

③ Think of 80 ÷ 20, and guess the quotient by 8 ÷ 2.

④ Is the quotient 4 correct? Let's confirm.

Division algorithm to calculate 84 ÷ 21 in vertical form

$21\overline{)84}$ ☐ → $2\square\overline{)8\square}$ 4 → $21\overline{)84}$ 4, 84 → $21\overline{)84}$ 4, 84, 0

From which place value → Divide → Multiply → Subtract

Summary By thinking which place value the quotient will be, we can calculate in the same way as we have learned.

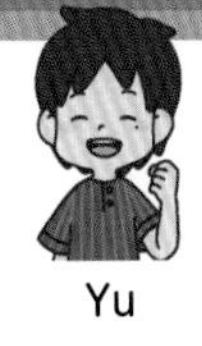
Yu

1 Let's calculate the following divisions in vertical form.

① 99 ÷ 33 ② 63 ÷ 21 ③ 64 ÷ 32

④ 48 ÷ 23 ⑤ 97 ÷ 32 ⑥ 91 ÷ 44

? Can you guess the quotient easily?

2 Let's think about the division algorithm to calculate $96 \div 33$ in vertical form.

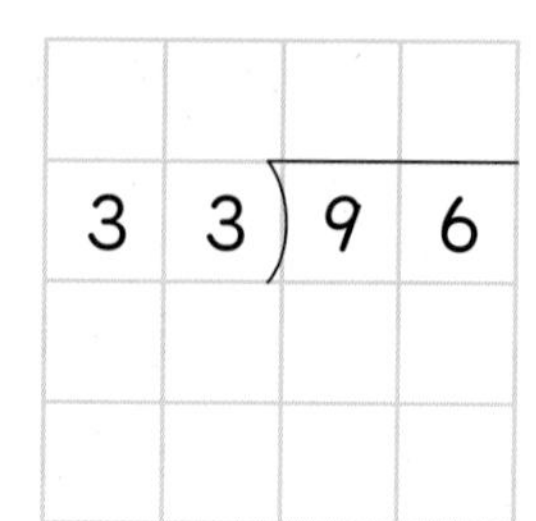

❶ Let's guess the quotient.

As for $90 \div 30$, if the quotient for $9 \div 3$ is 3...
Yu

❷ Let's try the calculation with quotient 3.

$$\begin{array}{r} 3 \\ 33\overline{)96} \\ 99 \end{array}$$

\ Want to know /

? Purpose What should we do in vertical form if the guess quotient is too large?

$33\overline{)96}$ → $3\overline{)9}$ → $\begin{array}{r} 3 \\ 33\overline{)96} \\ 99 \end{array}$ Cannot subtract. → Make the quotient smaller by 1 unit. → $\begin{array}{r} 2 \\ 33\overline{)96} \\ 66 \\ \hline 30 \end{array}$

30 is smaller than 33.

The first guess of the quotient is called tentative quotient. If the tentative quotient is too large, you have to replace it with a quotient that is smaller by 1 unit.

❸ Let's confirm the answer.

1 Let's guess the quotient and calculate. Then, let's confirm the answer.

① $56 \div 14$	② $60 \div 12$	③ $68 \div 24$
④ $94 \div 32$	⑤ $67 \div 23$	⑥ $79 \div 13$

? Is there a case that the quotient is too big even when I make the tentative quotient smaller by 1 unit?

Correcting the tentative quotient →

3 Let's think about how to calculate 68÷16 in vertical form.

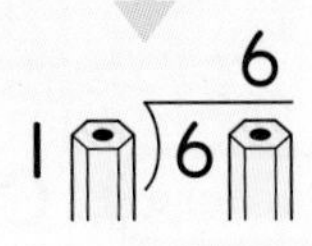

① Write a tentative quotient.

② Multiply the divisor and the tentative quotient.

But 16 × 6 = 96, so that's too big.

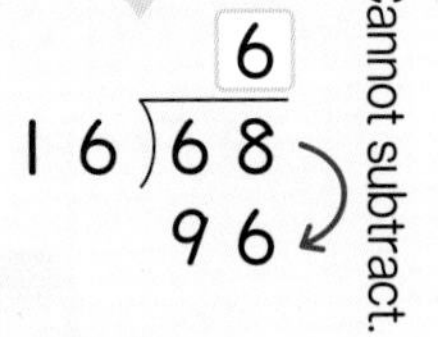

③ Replace it with a number that is smaller by 1 unit.

How about this time? 16 × 5 = 80 Still too big.

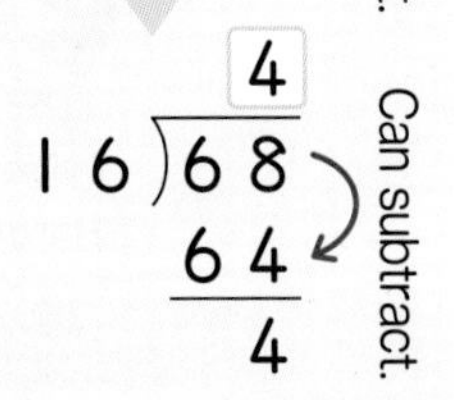

④ Make the tentative quotient smaller by 1 unit again.

I can. 4 is the correct quotient.

4
16)68
64
4
Can subtract.

1 Let's think about how to calculate 57 ÷ 18 by guessing a tentative quotient as shown above. Then, confirm the answer.

! Summary

If the tentative quotient you set originally is too large, replace it with a quotient that is smaller by 1 unit until the correct quotient is found.

2 Let's guess the quotient and calculate. Let's confirm the answer.

① 70 ÷ 14　② 69 ÷ 15　③ 97 ÷ 16

④ 72 ÷ 15　⑤ 53 ÷ 16　⑥ 82 ÷ 29

? Can we calculate in vertical form even when the dividend is a 3-digit number?

4 Let's think about how to calculate 170 ÷ 34 in vertical form.

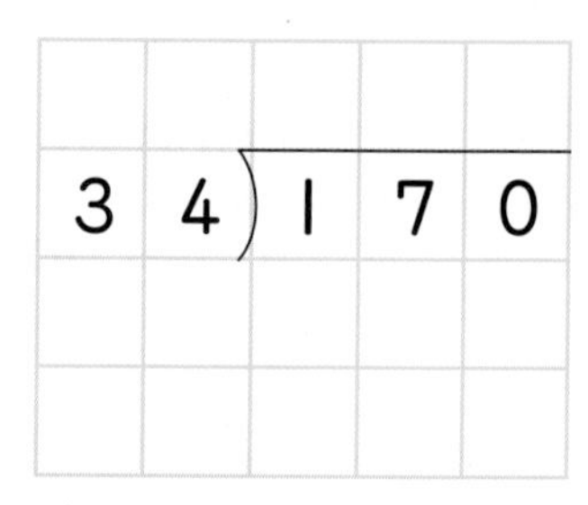

\ Want to think /

Purpose Can we divide 3-digit numbers using vertical form in the same way as we have learned before?

① From which place value can the quotient be written?

Sara

Since I cannot do 17 ÷ 34, the quotient cannot be written in the tens place.

② Let's guess the quotient.

Haruto

Think of 170 ÷ 30, the quotient for 17 ÷ 3 is...

3 4) 1 7 0

Division algorithm to calculate 170 ÷ 34 in vertical form

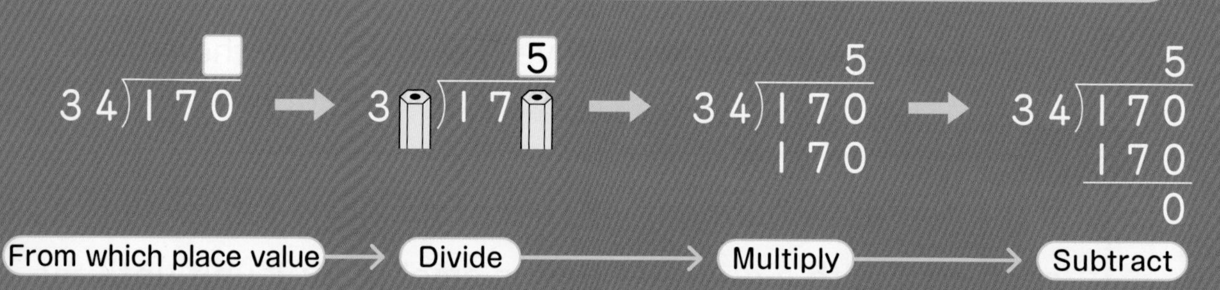

1 Let's think about how to calculate 229 ÷ 38 in vertical form.

3 8) 2 2 9

Akari

22 ÷ 3, so...I guess the quotient will be ...

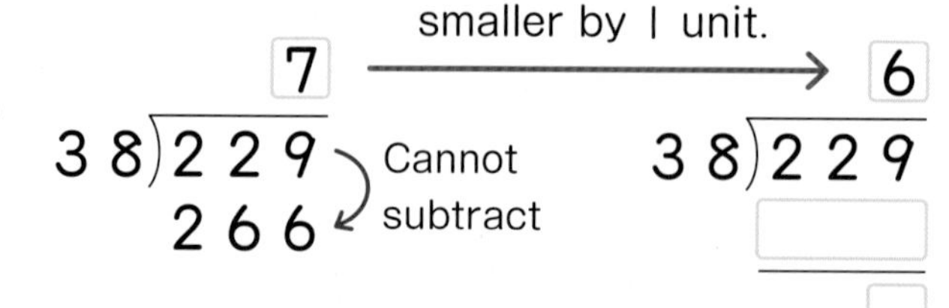

2 Let's think about how to calculate $326 \div 36$ in vertical form.

$36\overline{)326}$

① From which place value can the quotient be written?

② Let's guess the quotient.

Division algorithm to calculate 326 ÷ 36 in vertical form

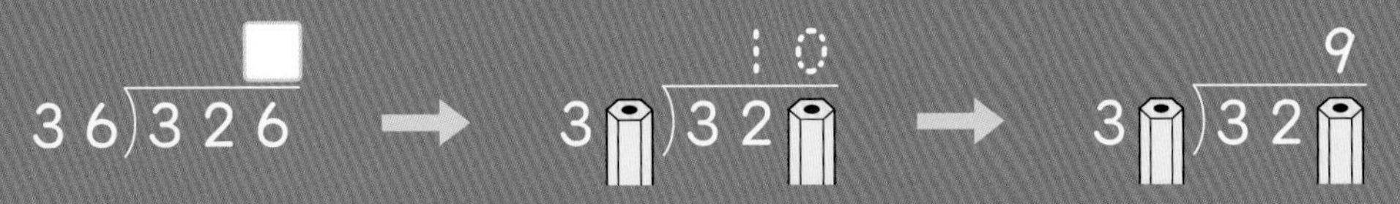

Summary

Even when the dividend has three digits, write the tentative quotient, and find out the correct quotient by reducing 1 unit.

3 Let's calculate the following in vertical form.

① $255 \div 51$	② $284 \div 71$	③ $201 \div 63$
④ $191 \div 24$	⑤ $365 \div 48$	⑥ $208 \div 21$
⑦ $711 \div 78$	⑧ $217 \div 25$	⑨ $257 \div 29$

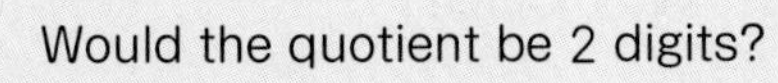

3 Division by 2-digit number (2)

1 We want to divide 252 sheets of colored paper so that 12 children will receive equally. How many sheets will each child receive?

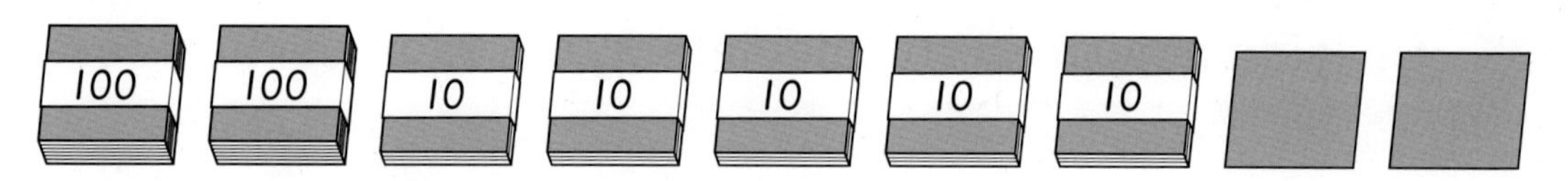

① Let's write a math expression. [ ]

② Let's make a guess on how much the quotient is.

③ From which place value can the quotient be written?

Sara

12 times 10 is 120, so the quotient seems to be larger than 10.

We can't divide bundles of 100 sheets to 12 children.

Haruto

Cannot write here.

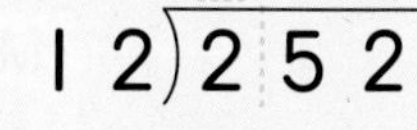

12)252

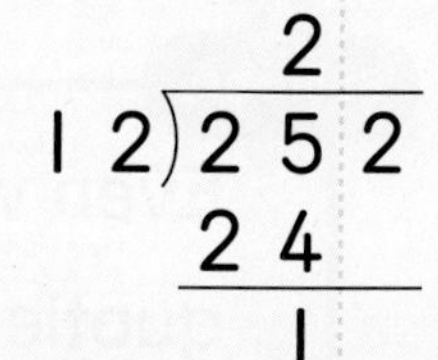

2
12)252
24
1

21
12)252
24
12
12
0

? **Purpose** ＼ Want to think ／

How can we calculate division in vertical form when the quotient is larger than 10?

④ Let's change the 100 sheet bundles into 10 sheet bundles and divide by 12 children.

$25 \div 12 =$ [ ] remainder [ ]

⑤ Let's change the remaining 10 sheet bundle into single sheets, add 2 single sheets, and divide by 12 children.

$12 \div 12 =$ [ ]

⑥ How many sheets of colored paper will each child receive?

Division algorithm to calculate 252 ÷ 12 in vertical form ▷

```
    □          2          2           2
12)252  →  12)252  →  12)252  →  12)252
                          24          24
                                      ──
                                       1
```

From which place value → Divide → Multiply → Subtract →

```
    2          21          21          21
12)252  →  12)252  →  12)252  →  12)252
   24          24          24          24
   ──          ──          ──          ──
    12          12          12          12
                            12          12
                                        ──
                                         0
```

→ Bring down → Divide → Multiply → Subtract

To divide in vertical form, after deciding and writing the quotient, repeat the calculation steps of "divde," "multiply," "subtract," "bring down."

1 Let's calculate $980 \div 28$ in vertical form.

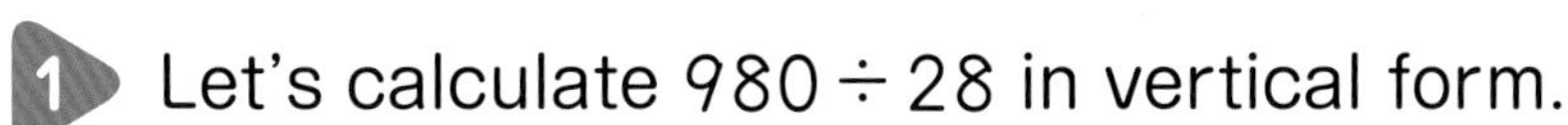

```
28)980
```

2 Let's calculate the following in vertical form.

① $736 \div 16$ ② $810 \div 18$
③ $432 \div 18$ ④ $851 \div 26$
⑤ $798 \div 35$ ⑥ $585 \div 39$
⑦ $612 \div 36$ ⑧ $578 \div 23$ ⑨ $939 \div 37$

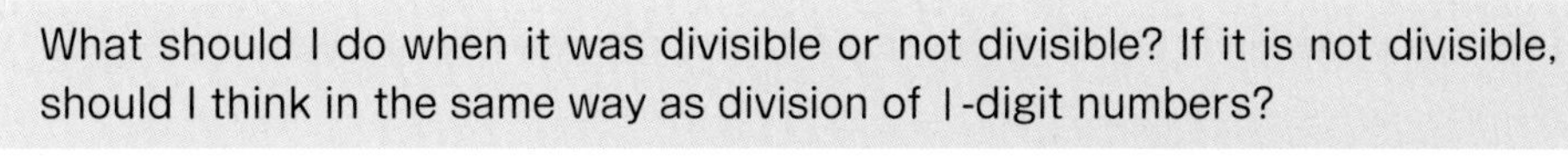

?
What should I do when it was divisible or not divisible? If it is not divisible, should I think in the same way as division of 1-digit numbers?

2 Let's think about how to calculate $607 \div 56$ in vertical form.

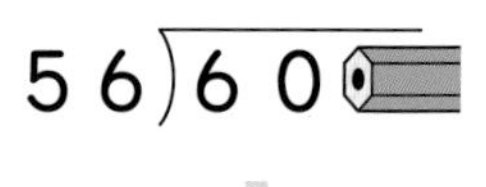

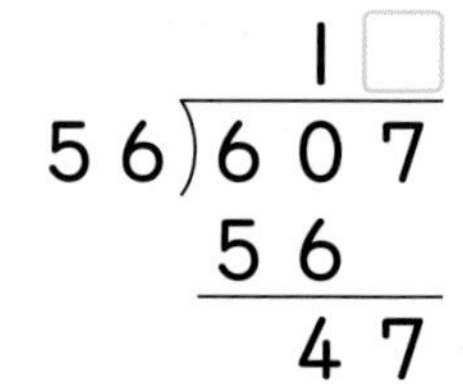

❶ From which place value is the quotient written?

❷ What number should be written in the ones place of the quotient?

Purpose What should I do if I could not divide in the calculation of the ones place?

1 $859 \div 21$ is calculated in vertical form. Let's explain ⓐ and ⓑ algorithms shown on the right.

ⓐ
```
     40
21)859
   84
    19
    00
    19
```

ⓑ
```
     40
21)859
   84
    19
```

Summary In case when we have 0 as a quotient, we need to omit the calculation for that process.

2 Let's calculate the following in vertical form.

① $705 \div 34$　② $913 \div 13$　③ $856 \div 42$

④ $531 \div 26$　⑤ $576 \div 56$　⑥ $942 \div 47$

⑦ $643 \div 16$　⑧ $750 \div 25$　⑨ $900 \div 18$

3 Let's correct the mistakes on the following calculations in vertical form.

①
```
      2
22)446
   44
     6
```

②
```
     21
31)645
   62
    25
    31
     6
```

③
```
     10
57)704
   57
    34
```

? Can we calculate in vertical form even when the dividend is a 3-digit number?

3 Let's think about how to calculate $942 \div 314$ in vertical form.

\ Want to think /

(Purpose) Even when both the dividend and the divisor are 3-digit numbers, can we calculate in the same way as we have learned?

Haruto

3	1	4) 9	4	2

1. From which place value is the quotient written?
2. Let's guess the quotient and calculate.

1 Let's calculate the following in vertical form.

① $744 \div 124$	② $928 \div 232$	③ $969 \div 323$
④ $639 \div 213$	⑤ $696 \div 348$	⑥ $952 \div 136$
⑦ $860 \div 213$	⑧ $524 \div 238$	⑨ $950 \div 289$

? We found various rules of division. Can we adapt them to large numbers?

That's it!

Division in various countries

The two calculations show division methods to calculate $984 \div 23$ in other countries. Let's compare the with the Japanese division algorithm in vertical form.

Canada

```
984)23
92   42
 64
 46
 18
```

Germany

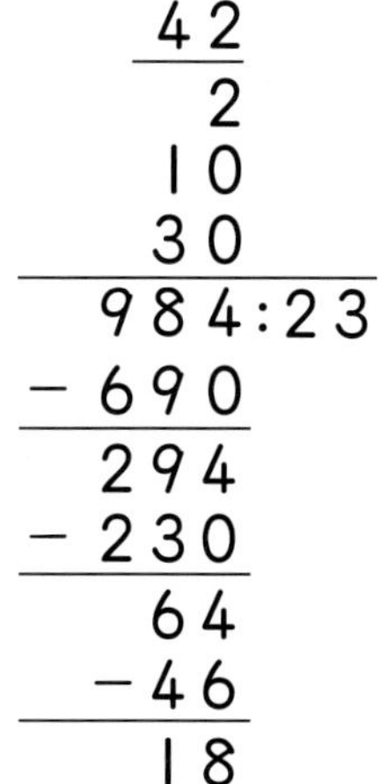

In Germany, they start with a small tentative quotient and repeat many times.

4 Rules of division

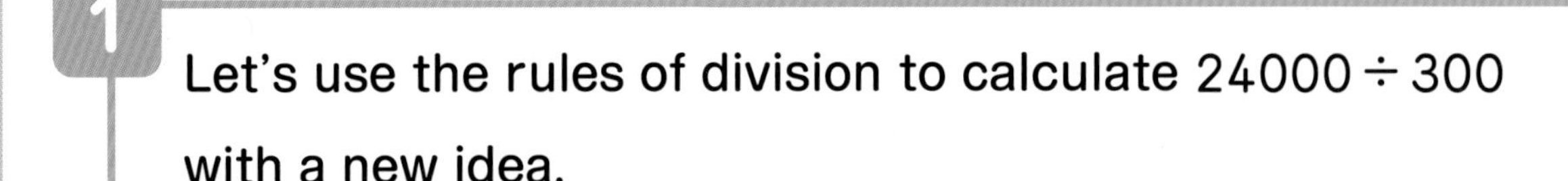

\ Want to think /

Purpose How can we use the rule of division?

Even if we divide the dividend and the divisor with the same number...

$24000 \div 300 = \square$

↓ ÷ 100 ↓ ÷ 100

$\square \div \square = \square$

Way to see and think

It is using the rule of the dividend and the divisor.

❶ Let's calculate in vertical form.

$$\begin{array}{r} 80 \\ 3\cancel{0}\cancel{0}\overline{)240\cancel{0}\cancel{0}} \\ \underline{24} \\ 0 \end{array}$$

1 Let's use the rules of division to calculate 3700 ÷ 500 with a new idea.

① Let's think about how to calculate.

② What is the remainder?

$$\begin{array}{r} 7 \\ 5\cancel{0}\cancel{0}\overline{)37\cancel{0}\cancel{0}} \\ \underline{35} \\ 2 \end{array}$$

$3700 \div 500 = \square$ remainder $\square$

③ Let's confirm the answer.

$500 \times \square + \square = \square$

Divisor Quotient Remainder Dividend

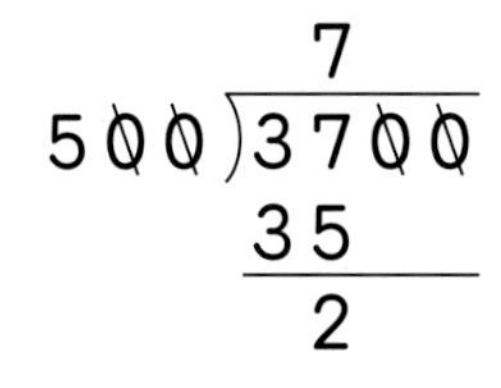

Summary We can use the rules of division, but we need to be careful about the amount of remainder.

Sara

2 Let's calculate the following in vertical form.

① 1700 ÷ 800 ② 4400 ÷ 600 ③ 7500 ÷ 80

? In what problems can division be used?

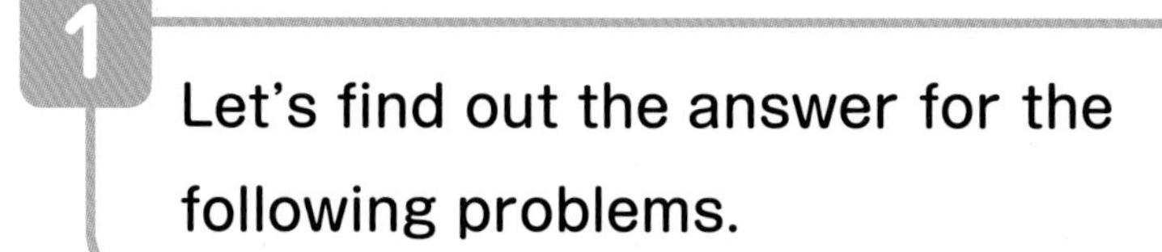

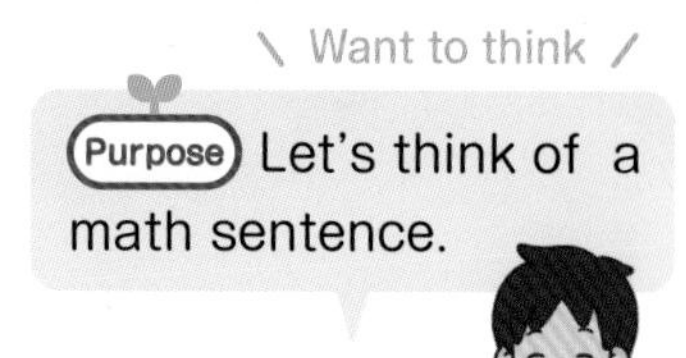

Want to think
Purpose Let's think of a math sentence.
Yu

1 Let's find out the answer for the following problems.

① We want to distribute 18 sheets of colored paper to each of 7 groups. How many sheets are needed?

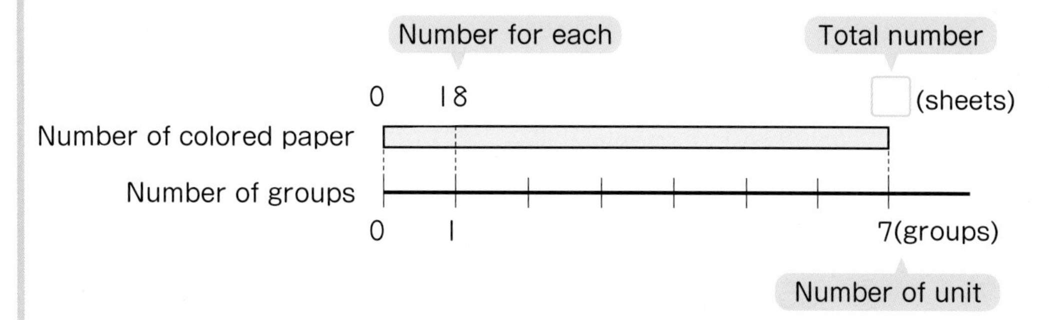

② We want to distribute 126 sheets of colored paper so that each group receive 18 sheets. How many groups can receive?

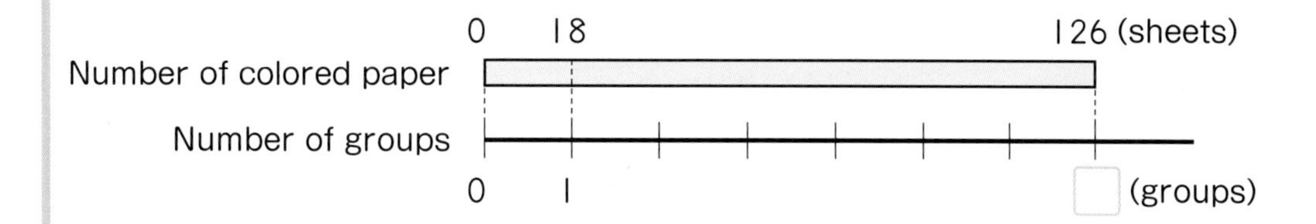

③ There are 126 sheets of colored papers. If we distribute equally to 7 groups, how many sheets will each group receive?

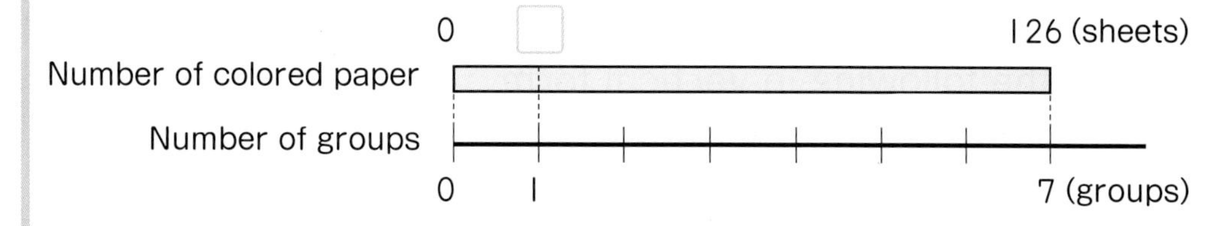

What we want to find out is different among the three problems.

Summary We can use multiplication to find out the total number. We can use division to find out the number for each and number of unit.

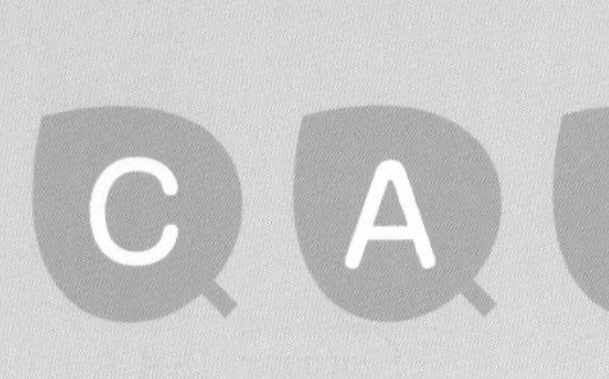

What can you do? 

☐ We understand how to calculate division by 2-digit numbers. → p.96

1 Let's summarize how to calculate division by 2-digit numbers.

① As for the calculation on the right, the quotient will be written on the ☐ place. $32\overline{)768}$

② The quotient of the tens place will be set by ☐ ÷ ☐.

③ The calculation to find out the quotient of the ones place is ☐ ÷ 32.

☐ We can solve a division with a 2-digit divisor. → pp.89 ~ 98

2 Let's calculate the following in vertical form.

① 60 ÷ 20　② 140 ÷ 70　③ 130 ÷ 40　④ 96 ÷ 32

⑤ 97 ÷ 27　⑥ 85 ÷ 19　⑦ 344 ÷ 43　⑧ 385 ÷ 56

⑨ 411 ÷ 45　⑩ 672 ÷ 28　⑪ 453 ÷ 17　⑫ 738 ÷ 24

☐ We can make a division expression and find out the answer. → pp.89 ~ 98

3 Let's answer the following questions.

① If you buy some bread, each piece will cost 75 yen. If the total amount paid is 900 yen, how many pieces of bread will you get?

② 342 screws are distributed to bags in groups of 18 screws. How many bags are needed?

③ 800 stickers are distributed to 12 children. How many will each child receive, and what would the remainder be?

☐ We understand the rules of division. → p.100

4 Let's calculate the following in vertical form.

① 80000 ÷ 200　② 3200 ÷ 160

Supplementary Problems → p.148

Which "Way to See and Think Monsters" did you find in "8 Division by 2-digit Numbers"?

I found "Unit" when I was trying to calculate division in large numbers.

When I was confirming the answer...

Utilize Usefulness and Efficiency of Learning

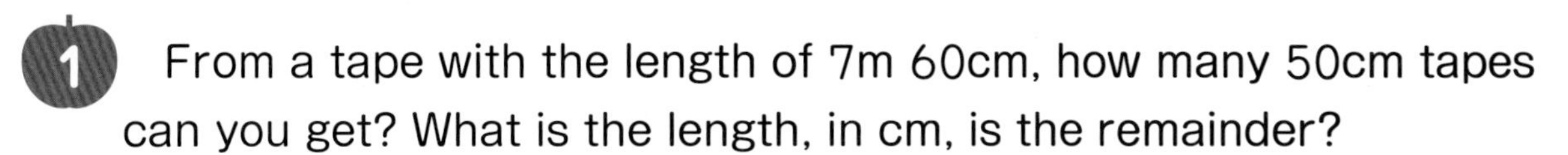

1 From a tape with the length of 7m 60cm, how many 50cm tapes can you get? What is the length, in cm, is the remainder?

2 There are 38 students in Kaori's class. When 570 sheets of origami are divided, how many sheets of origami will each child receive? If there is a remainder, also answer it.

3 Let's explain the reasons why $320 \div 40$ can be calculated by $32 \div 4$.

4 Read through ⓐ to ⓕ and answer the following questions.

ⓐ We want to use eight tapes of 160cm length. How many cm will you need in total?
ⓑ We divided 160 sheets of colored paper to children. The remainder was 8 sheets. How many sheets did we have in total?
ⓒ There are 160 candies. If we distribute 8 to each child, how many children can receive the candies?
ⓓ A boy has 160 cards. He gave 8 cards to his brother. How many cards does the boy have now?
ⓔ A tape costs 160 yen for 8 m. How much is this tape for 1 m?
ⓕ There are 160 children. If we distribute 8 candies to each of them, how many candies are needed?

① Which of the above questions will be expressed by the math expression $160 \div 8$?
② Which of the above questions will be expressed by the math expression 160×8?

5 In the table on the right, the vertical, horizontal, and diagonal products have the same answer. Let's fill in each empty space with a number.

12	ⓐ	2
ⓑ	6	36
18	ⓒ	ⓓ

With the Way to See and Think Monsters...

Let's Reflect!

Let's reflect on which monster you used while learning "8 Division by 2-digit Numbers."

Unit

By setting 10 as one unit, we could calculate in the same way as the division of 1-digit numbers.

① In what way did you calculate $80 \div 20$?

Setting 10 as one unit, I found out the answer by doing 8 ÷ 2.
Yu

Why

We could explain why the answer or the remainder is not correct by confirming the answer.

② What did you pay attention to in the calculation of $80 \div 30$?

I set 10 as one unit, and did 8 ÷ 3.
Akari

8 ÷ 3 makes the remainder 2, but since we assumed 10 to be one unit, we had to consider the remainder to be two 10s as well.
Haruto

We could confirm the answer by trying divisor × quotient + remainder=dividend.
Sara

Solve the ?

Division by 2-digit numbers could be done in the same way as before by using the rules of division and the rules of vertical form.

Sara

Want to connect

Can we calculate division in vertical form even when the numbers get large?

Haruto

How many times (1)

Distance of jump

1 A dolphin jumped to a high of 560cm.
The body length of this dolphin is 280cm.
Let's compare its distance of jump and the body length of the dolphin.

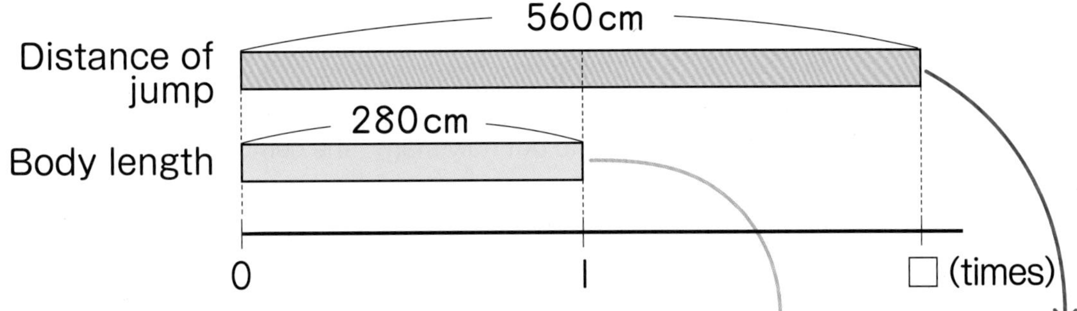

① Let's make a table to find out the relationship between the distance of the jump and the body length.

280cm	560cm
1 time	□ times

② Using ①, let's think of a way to find out how many times of the dolphin's body length it jumped.

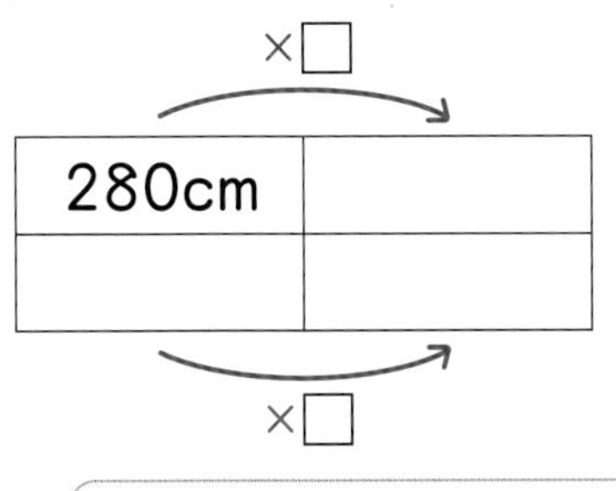

280cm	

\ Want to think /

Purpose How did we calculate to find out how many times of the base length the compared length is?

Yu: It is □ times, so we can express the math expression using □.

Haruto: In this case, "the base length" is the body length of the dolphin, and the "compared length" is the distance of jump.

Akari

③ How many times of its body length did the dolphin jump?

④ If we consider 280cm as 1 how many is 560cm? Look at the diagram on the right and think.

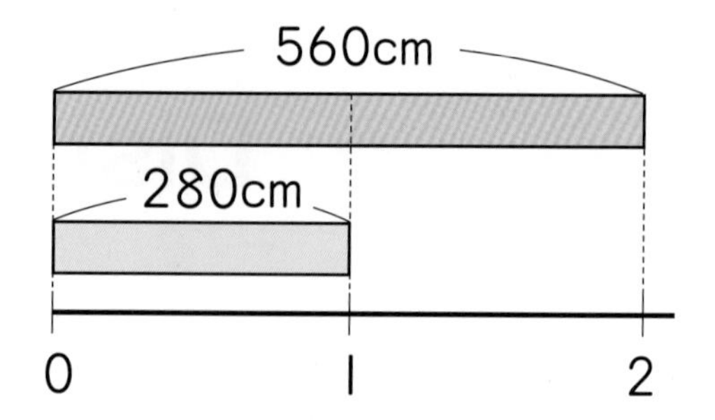

560cm is 2 times 280cm. So, if we consider 280cm as 1, 560cm can be represented as 2.

Summary We can find out by (compared length) ÷ (base length) = (how many times). If we know how many times, we can consider "the base length" as 1, we can find out how many "the compared length" will be.

1 An impala jumped 9m80cm. The body length of this impala is 140cm. Let's compare its distance of jump and the body length of the impala.

① Let's make a table to find out the relationship between the distance of the jump and the body length.

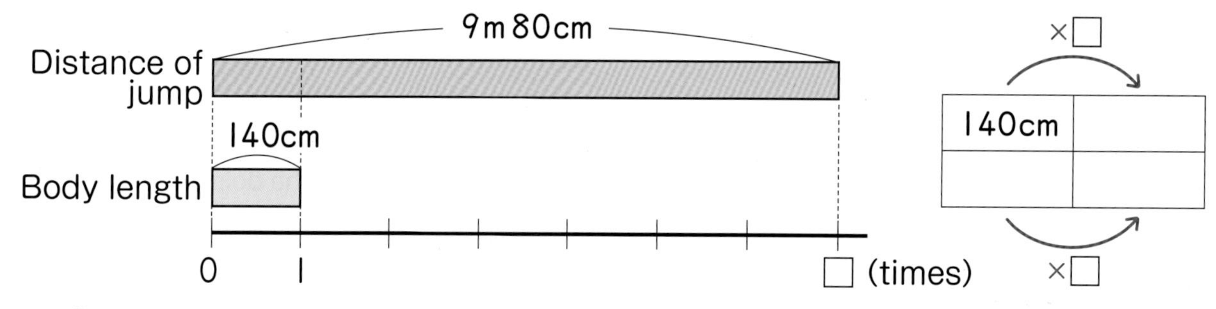

② How many times of its body length did the impala jump?

③ If we consider 140cm as 1, how many is 9m80cm?

2 There is a kangaroo that can jump 6 times its body length. Let's think about the statement below.

① If the body length of the kangaroo is 120cm, how many cm would it jump?

We need to find out the compared length.

Haruto

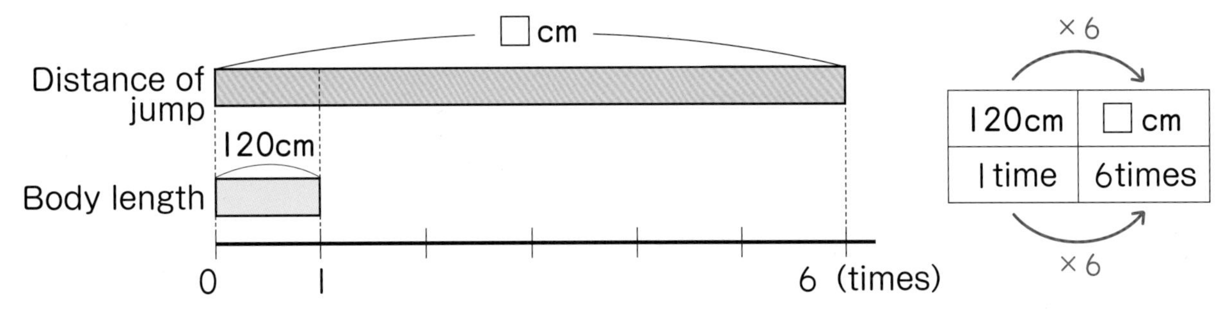

② If we consider 120cm as 1, how many cm would 6 represent?

3 There is a frog that jumps 40 times its body length. Let's think about the statement below.

① This frog can jump 3m20cm. What is the body length?

We need to find out the base length.

Akari

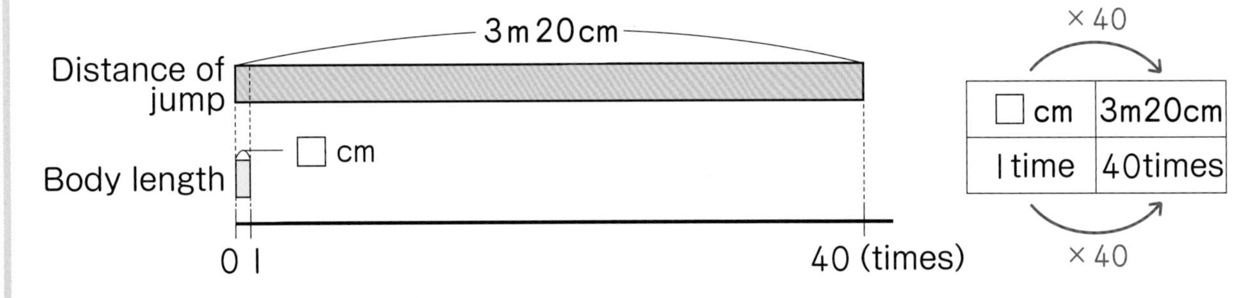

② If we set 3m20cm as 40, how many cm would 1 represent?

Let's solve math problems by using diagrams and tables.

When we don't know the total number

There are 16 candies in one bag. How many candies will there be with 14 bags?

Haruto: Let's draw a diagram and make a math sentence.

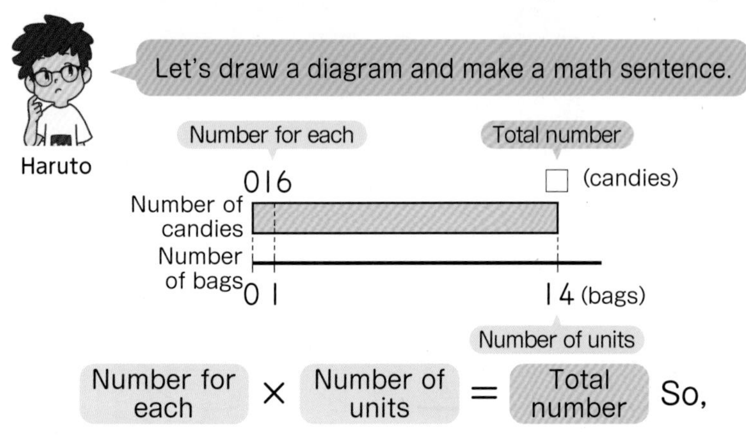

Number for each × Number of units = Total number So,

$16 \times 14 = 224$ Answer: 224 candies

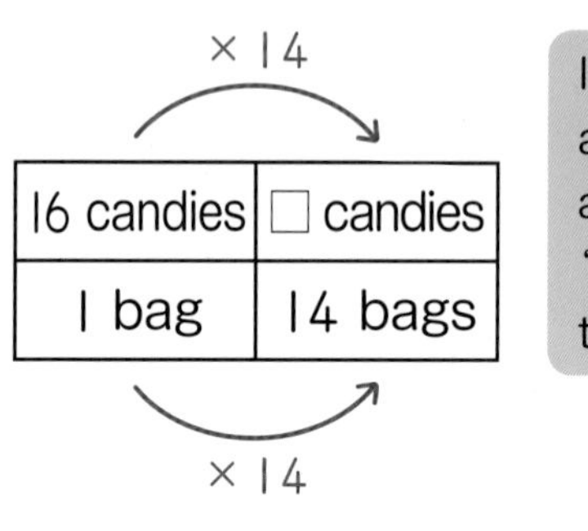

16 candies	☐ candies
1 bag	14 bags

Sara: If we draw a table and think "how many times"...

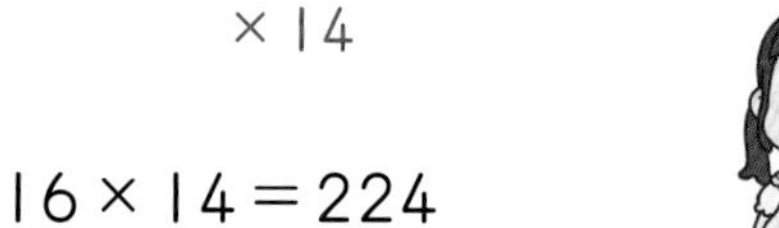

$16 \times 14 = 224$

Answer: 224 candies

How to draw a diagram ▷

(1) Draw a tape representing the number of candies and a straight line representing the number of bags.

0 (candies)
Number of candies
Number of bags
0 (bags)

(2) Draw a scale that represents 1 bag, and connect it with the scale that represents 16 candies.

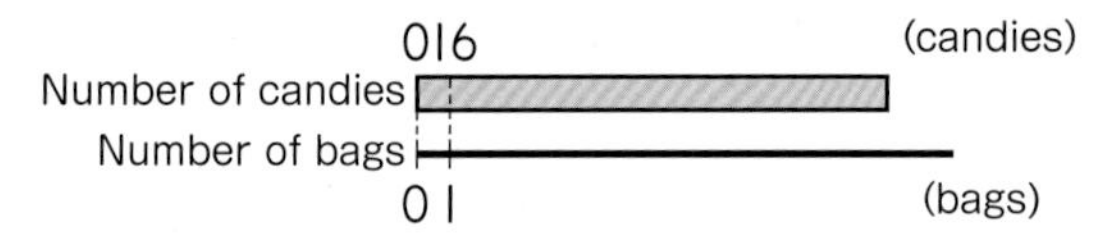

(3) Draw a scale that represents 14 bags, and draw a ☐ on the scale responding to it.

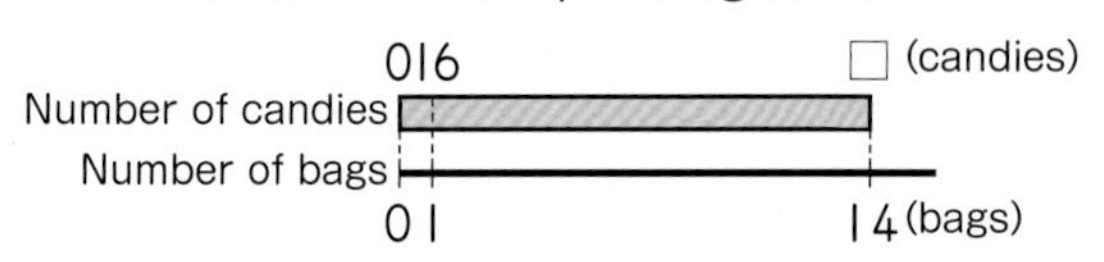

How to make a 4-cell table ▷

(1) Draw a table with 4 entries.

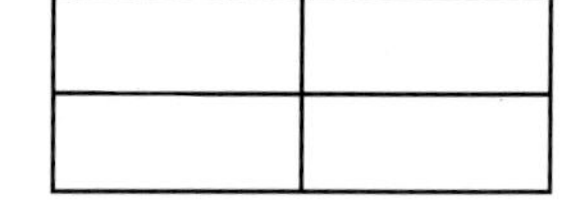

(2) Since 1 bag contains 16 candies, write "1 bag" and "16 candies" in the left column.

16 candies	
1 bag	

(3) Since we don't know the number of candies for 14 bags, write "14 bags" and "☐ candies" in the right column.

16 candies	☐ candies
1 bag	14 bags

You can also draw it based on the order of the math problem as shown on the right.

1 bag	14 bags
16 candies	☐ candies

The same unit is placed in the horizontal row.

In mathematics, it is sometimes easier to understand the situation of the problem represented by a diagram or a table. Let's review the various diagrams and tables we have learned so far.

When we don't know the number for each

There are 432 sheets of colored paper. If we distribute to 18 children equally, how many sheets will each child get?

Using a diagram···

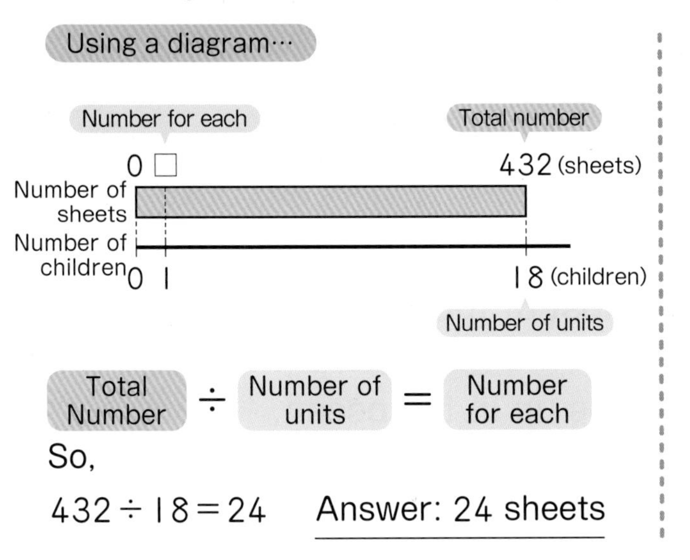

Total Number ÷ Number of units = Number for each

So,

$432 \div 18 = 24$ Answer: 24 sheets

Using a table···

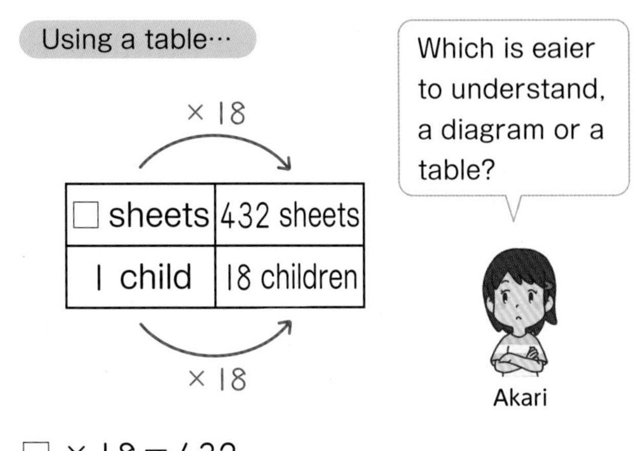

□ sheets	432 sheets
1 child	18 children

$\square \times 18 = 432$

$432 \div 18 = 24$ Answer: 24 sheets

When we don't know the number of units

We want to divide 108 pencils so that each child will get 12. How many children can share the pencils?

Using a diagram···

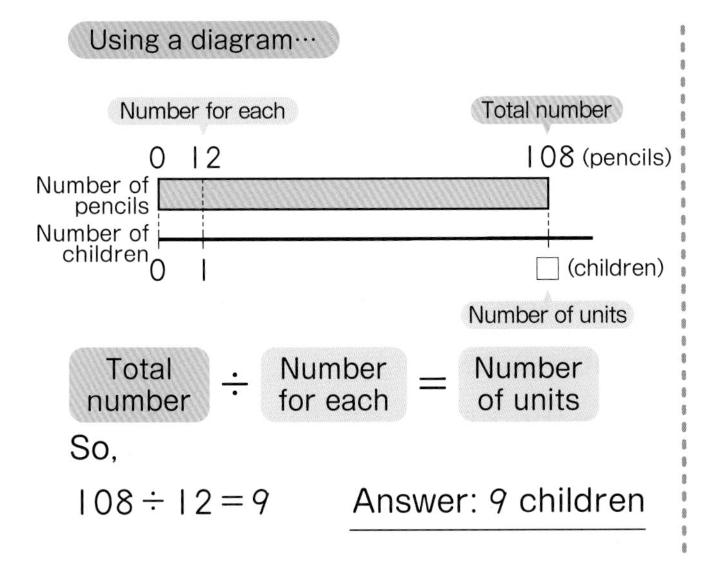

Total number ÷ Number for each = Number of units

So,

$108 \div 12 = 9$ Answer: 9 children

Using a table···

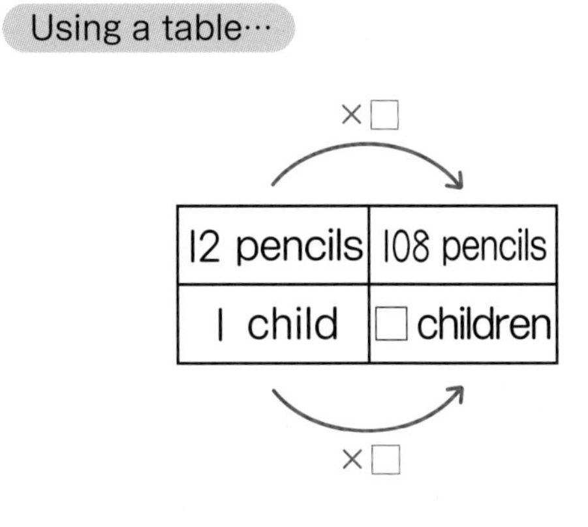

12 pencils	108 pencils
1 child	□ children

$12 \times \square = 108$

$108 \div 12 = 9$ Answer: 9 children

Utilizing Math for SDGs

Let's think about plastic waste

From July 2020, plastic bags are charged and the number of people carrying their own bags has increased. This is due to the problem of plastic waste.
Plastic is used in many places because it is light and strong. However, most of them are disposed of as waste, and some of them end up in the garbage, which passes through rivers and eventually ends up in the ocean.
As a result, there is about 150 million tons of plastic waste in the ocean today. In the world, about 8 million tons of plastic waste, (equivalent to 50,000 jet airplanes), continues to increase every year and Japan is thought to be responsible for 20,000 to 60,000 tons of this amount.
If plastic waste continues to grow at its current rate, by 2050 there will be more plastic waste in the oceans than fish.
What can we do to prevent this?

① Let's find out what kind of things end up in the ocean as plastic waste.

② The graph on the right page shows the amount of plastic produced, discarded as waste, and reused.
Let's discuss what you notice from the graph.

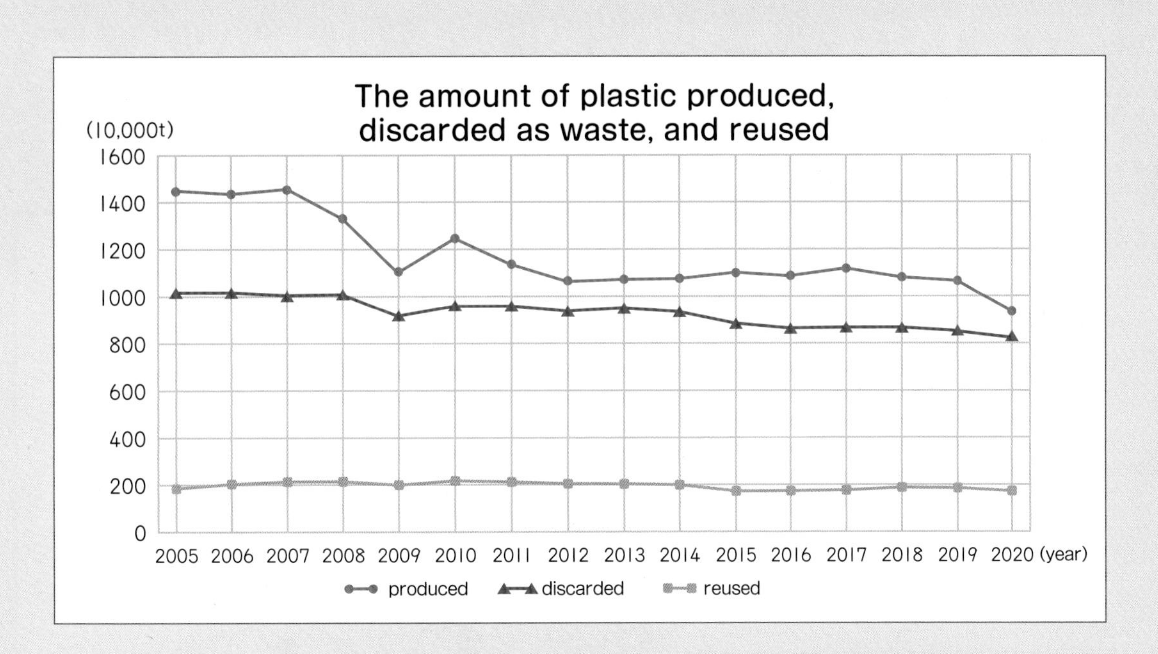

③ Let's discuss what you actually do and what we can do to reduce plastic waste.

Think back on what you felt through this activity, and put a circle.

Let's reflect on yourself!

	Strongly agree	Agree	Don't agree
① I could search for ways to reduce plastic waste.			
② I could read the graph and think about the amount of plastic waste.			
③ I could utilize the knowledge of mathematics when I investigated the amount of plastic waste.			

	Strongly agree
④ I am proud of myself because I did my best.	

Let's praise yourself with some positive words for trying hard to learn!

What kind of figure is hidden?

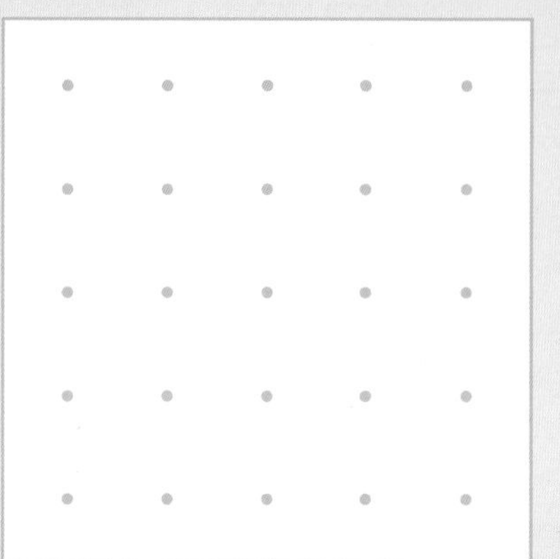

Let's use the card on page 157 and make various quadrilaterals by connecting the dots with four straight lines.
1

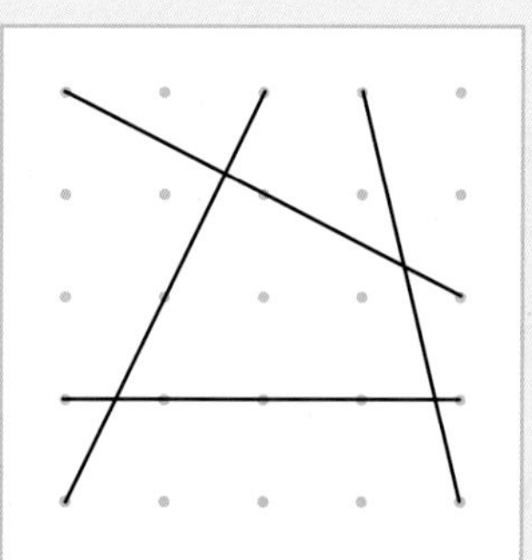

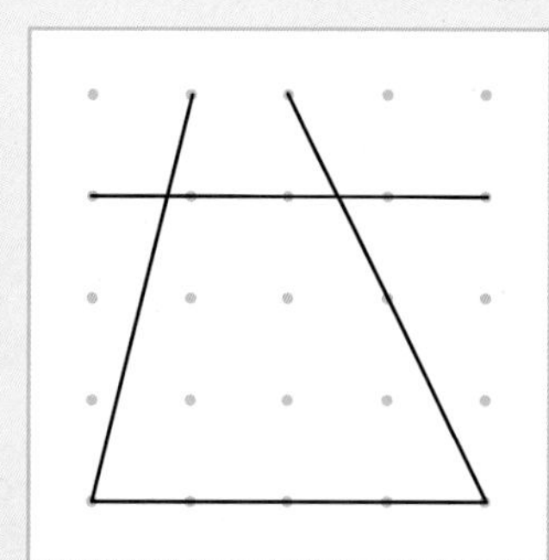

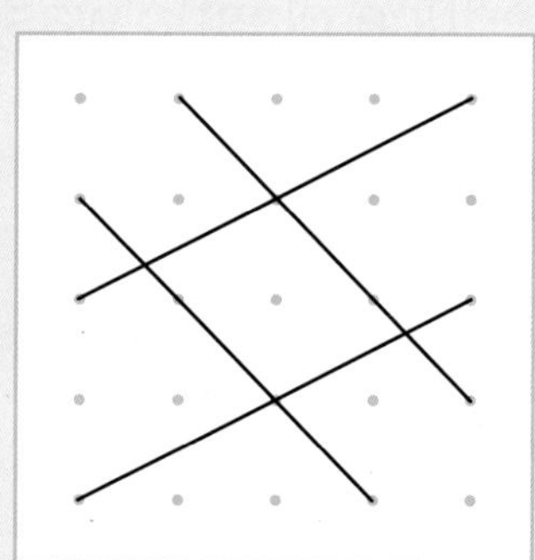

We can make various quadrilaterals.

Some straight lines are aligned, and others aren't.

We can make other quadrilaterals too.

Can we classify depending on their properties?
2

\ Want to try /

What are the similarities among the quadrilaterals you drew?

Let's explore the properties of quadrilaterals and classify them.

1 We made the following quadrilaterals in the previous page.
Let's find out the similarities among them.

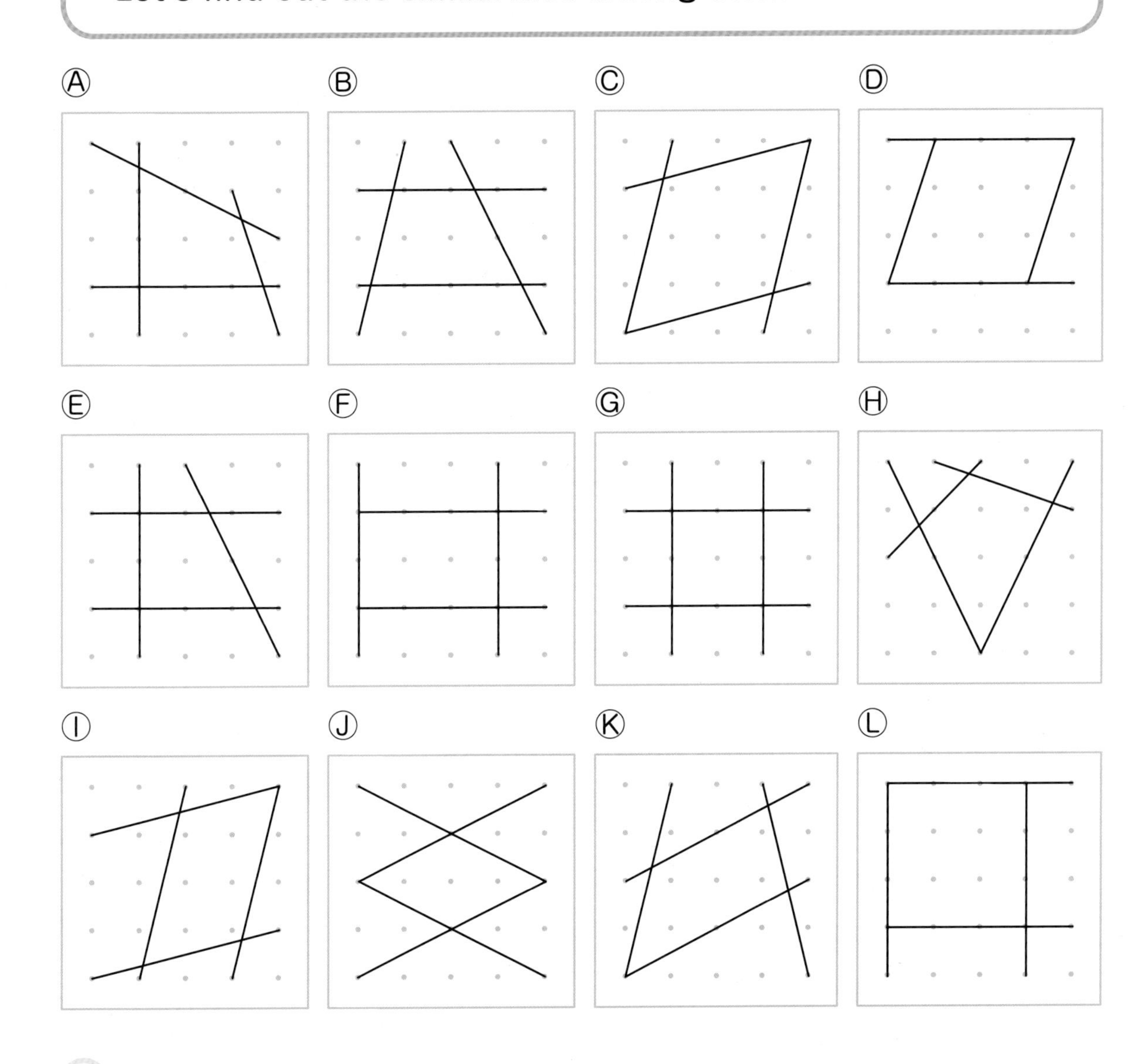

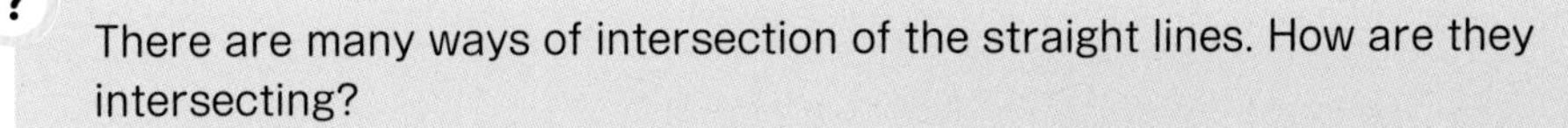
? There are many ways of intersection of the straight lines. How are they intersecting?

1 Perpendicular

1 Let's examine the quadrilateral Ⓔ on page 113.

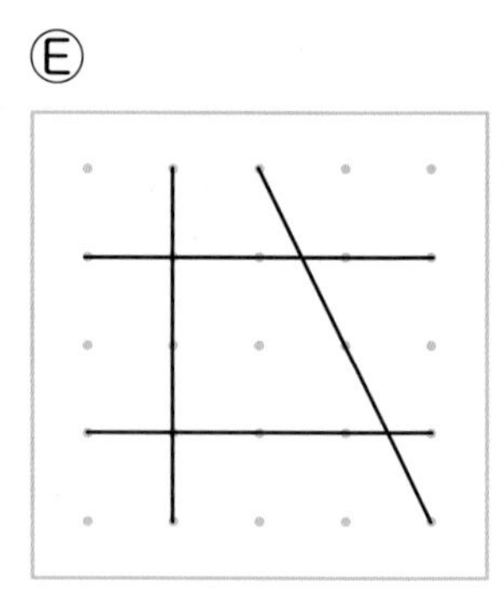

❶ At what angles do the straight lines (1) and (4) intersect?

❷ At what degrees do the straight lines (2) and (3) intersect? What are the size of angles ⓔ, ⓕ, ⓖ, ⓗ?

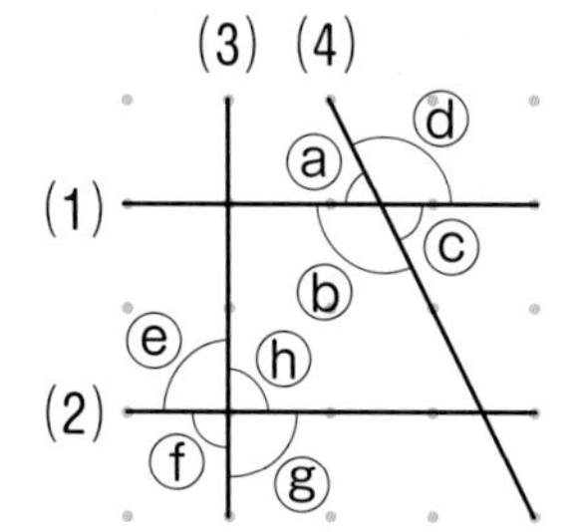

When the intersecting angle of two straight lines is a right angle, those two straight lines are **perpendicular** lines.

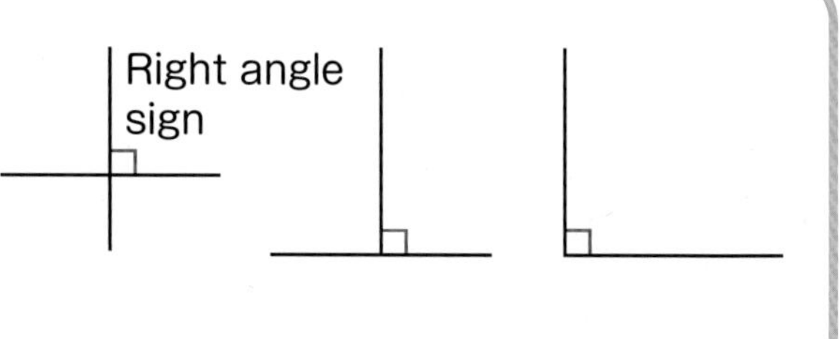

The two straight lines (2) and (3) are perpendicular lines.

Want to explore

? Purpose Can we find other perpendicular straight lines?

1 The diagram on the right shows the symbol for a post office in a Japanese map. Let's explore straight lines *a*, *b*, and *c*.

① At what angle do the straight lines *b* and *c* intersect?

② If you extend straight line *c*, how does it intersect with straight line *a*?

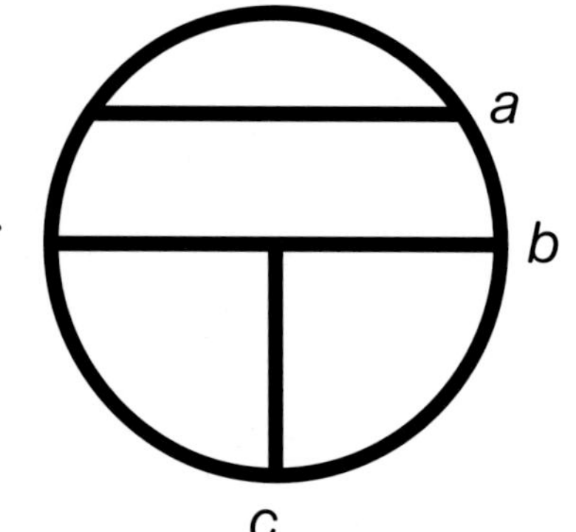

2 Let's find the places where the two straight lines intersect as perpendicular lines in the quadrilaterals on page 113.

3 Which two straight lines are perpendicular lines in the following diagrams?

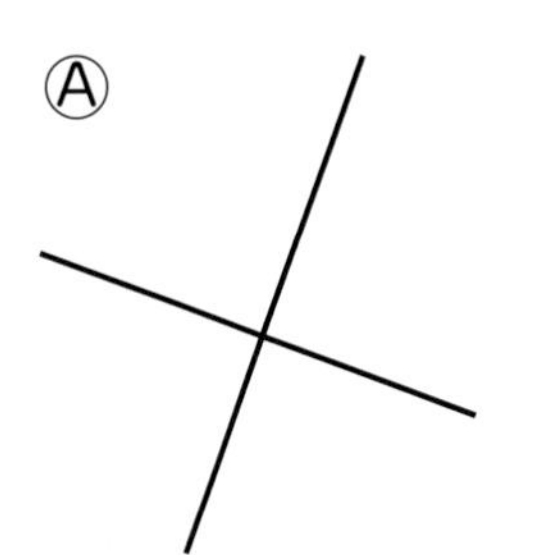

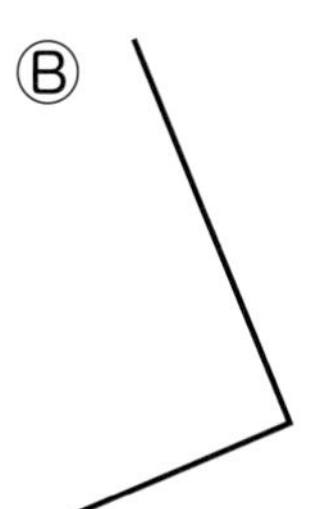

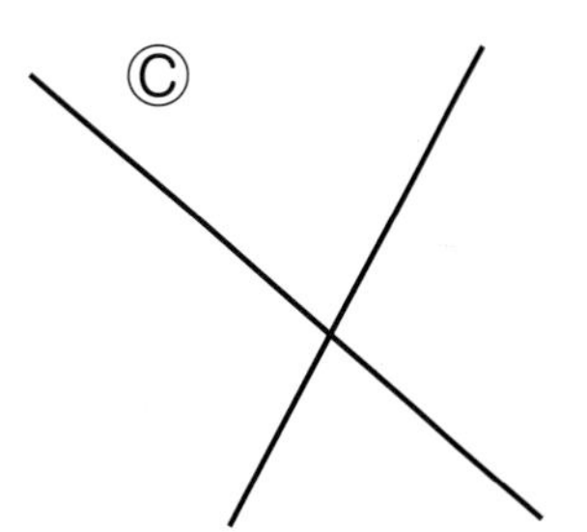

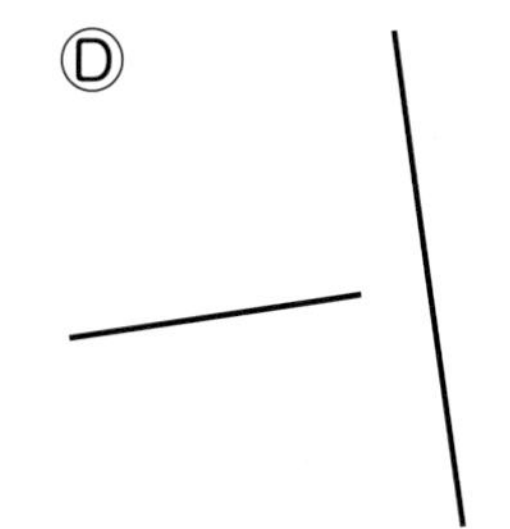

Even if two straight lines do not intersect, when one or both straight lines are extended and intersect in a right angle, they are called perpendicular lines.

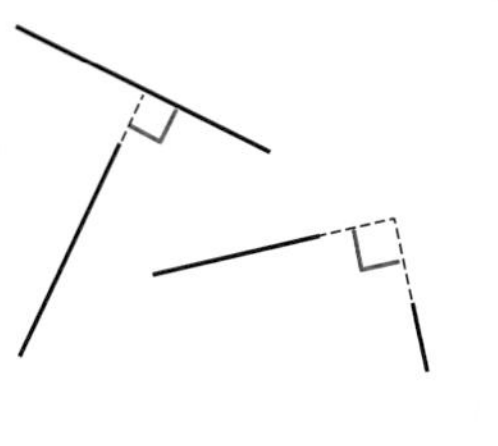

4 Let's fold a paper and make two straight lines that intersect as perpendicular lines.

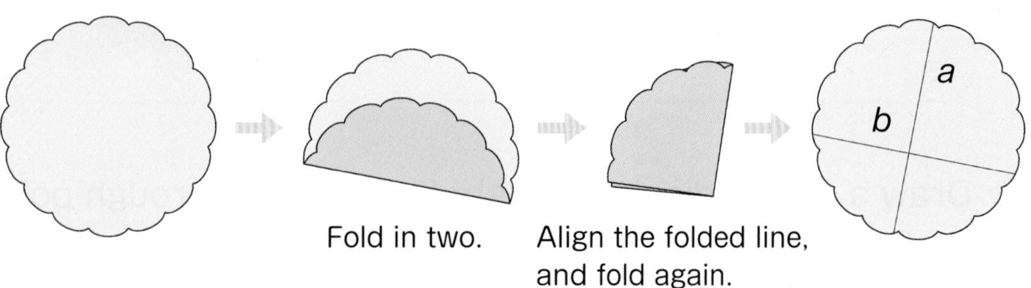

Fold in two.

Align the folded line, and fold again.

? Can we draw two perpendicular lines without connecting the dots or folding papers?

【垂】Dangle.

2

Akari and Yu drew a perpendicular straight line to one straight line. Let's explain their ways of drawing.

\ Want to try /

Purpose Let's draw a perpendicular straight line to a straight line.

Haruto

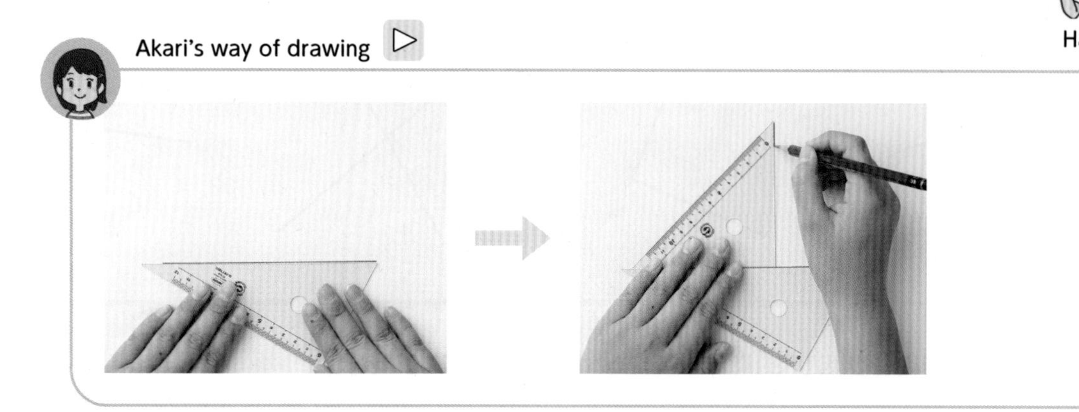

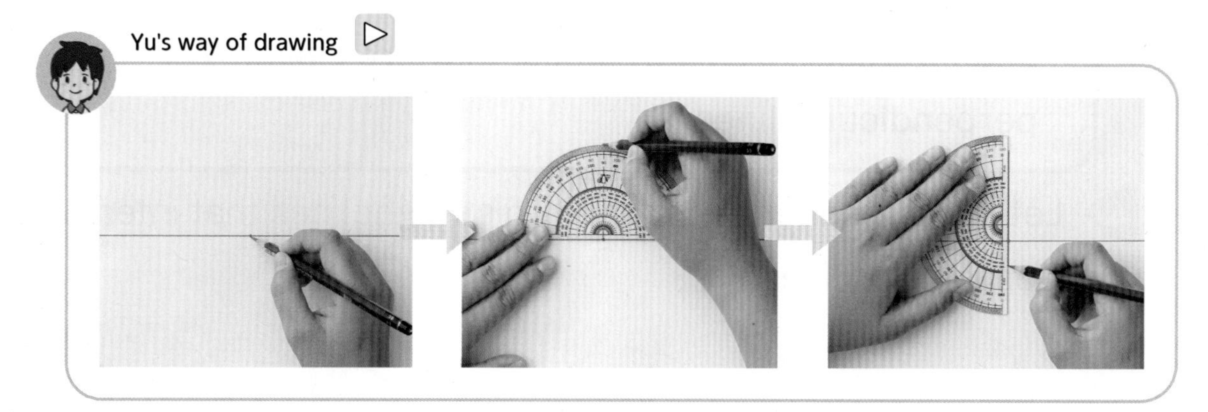

1 Draw a perpendicular straight line passing through point A to straight line *a*.

A •

a ———————————————————

How to draw a perpendicular straight line

When it goes through a point that is not on the straight line 

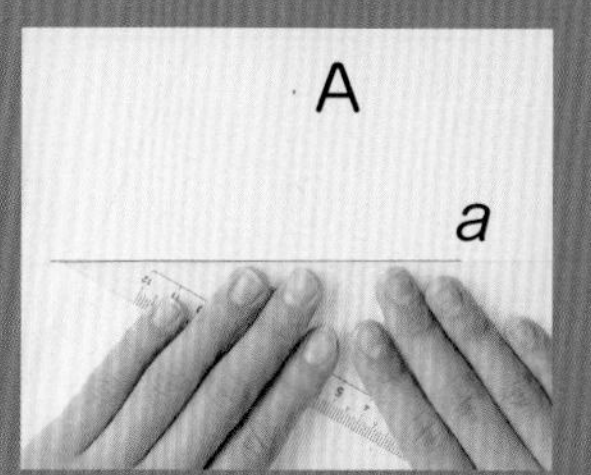

(1) Match the triangle ruler on straight line *a*.

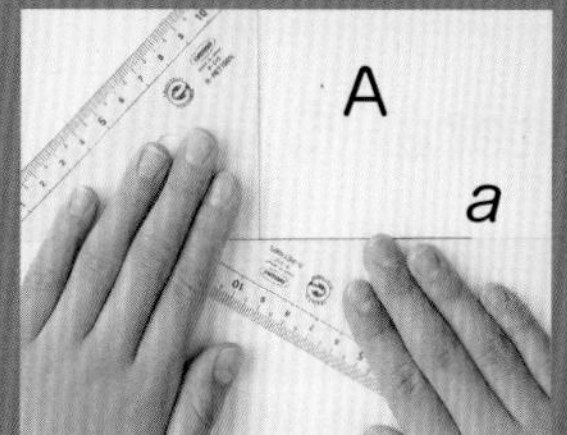

(2) Match the other triangle ruler's right angle side with straight line *a*.

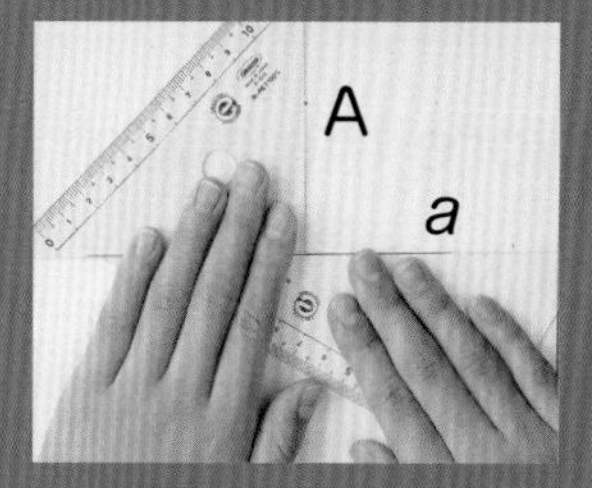

(3) Move the triangle rular above to match point A.

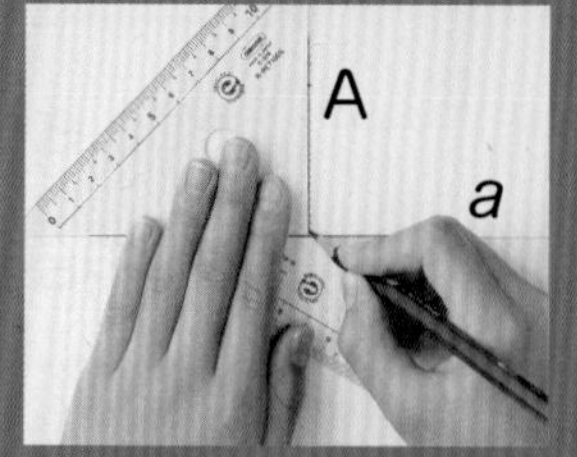

(4) Holding the triangle ruler, draw a straight line passing through point A.

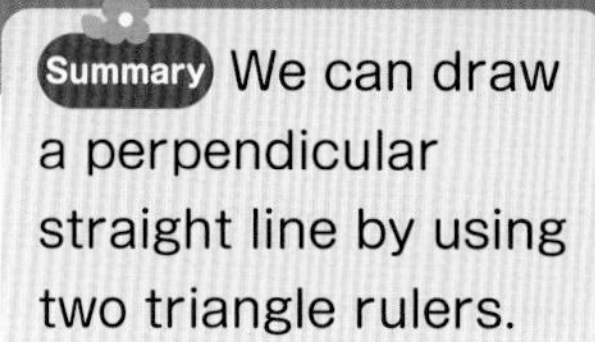

Summary We can draw a perpendicular straight line by using two triangle rulers.

Sara

2 Let's draw perpendicular straight lines to straight line *b*, through points B and C, respectively.

• B

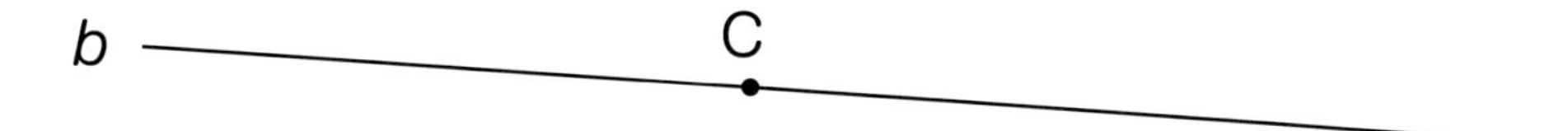

2 Parallel

1 Let's explore the quadrilateral Ⓔ on page 113.

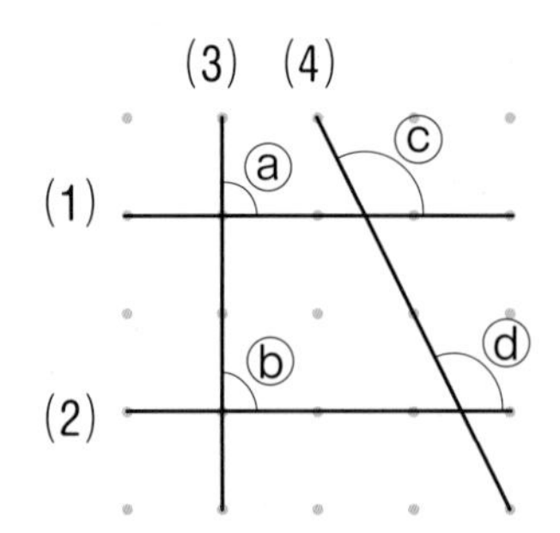

❶ How do straight lines (1) and (2) intersect straight line (3)?

Two straight lines are called **parallel** when a third straight line is perpendicular to both straight lines.

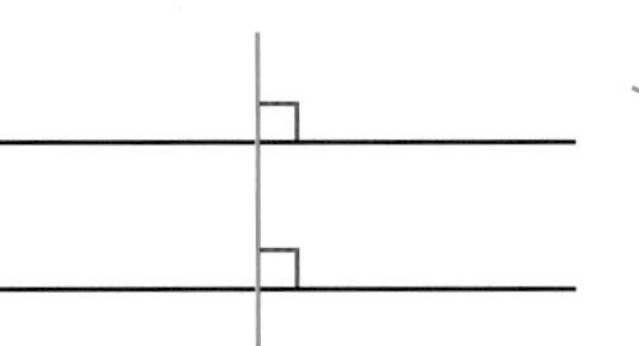

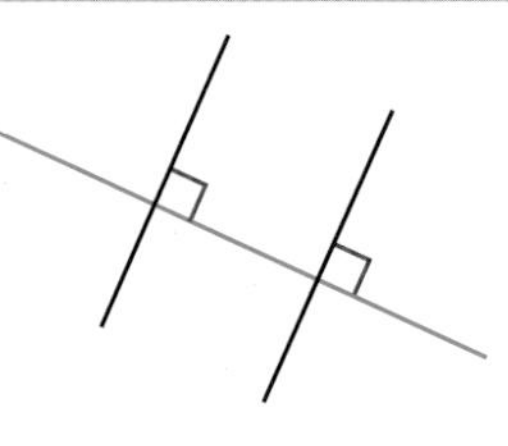

The straight lines above, (1) and (2), are parallel straight lines.

? Purpose — Want to try

At what size of the angle does a straight line and the parallel two straight lines intersect with each line?

❷ Let's compare the size of angles of ⓒ and ⓓ.

1 Straight lines *a* and *b* are parallel in the right diagram. Let's draw a diagonal straight lines and examine the size of angles formed by the intersection to straight lines *a* and *b*.

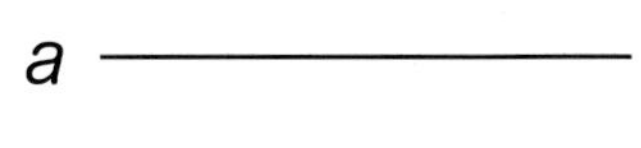

! **Summary**

Parallel straight lines are intersected by another straight line at an equal angle.

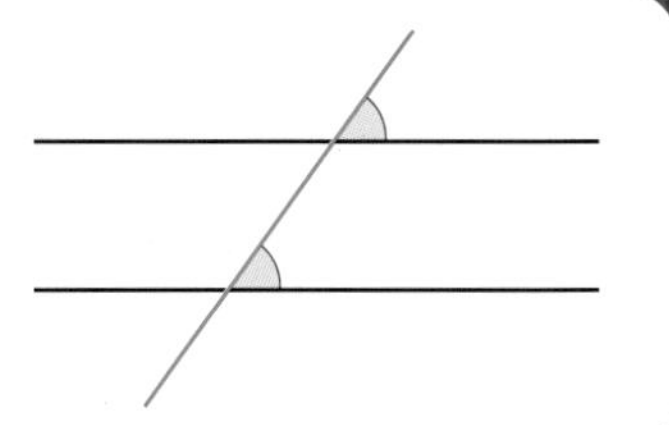

2 Are straight lines c and d parallel? Let's extend straight line c and examine.

d
c

The two straight lines are also parallel in the diagram on the right.

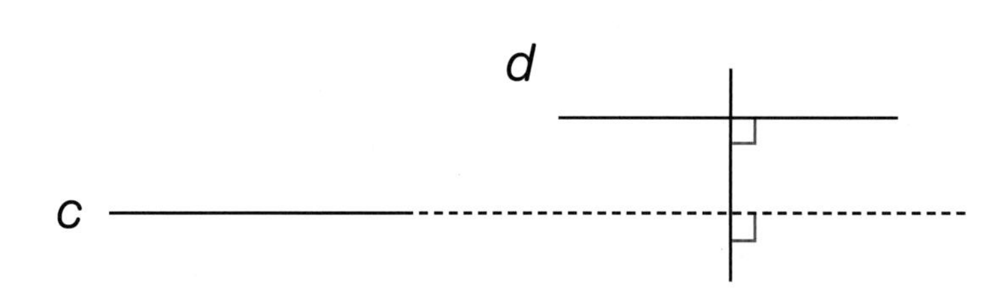

3 Let's find out the parallel straight lines in the quadrilateral on page 113.

4 Which straight lines are parallel in the following diagram?

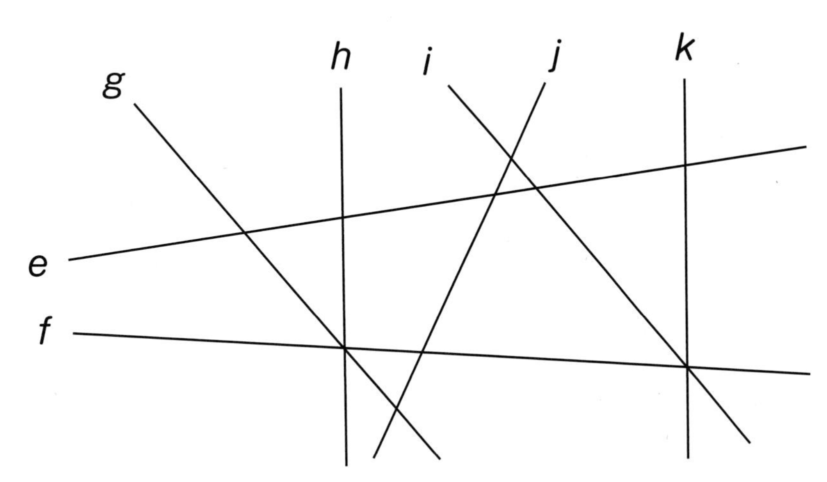

5 The straight lines l and m are parallel in the diagram on the right. Let's find out the size of angles for angles ⓔ to ⓖ.

n
l
110°
ⓖ
m
ⓕ
ⓔ

Let's try it without using a protractor.

?

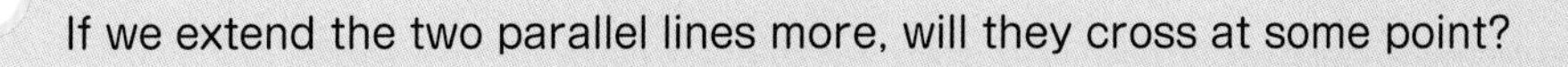

If we extend the two parallel lines more, will they cross at some point?

2 Straight lines *a* and *b* are parallel in the diagram on the right. Let's examine the followings.

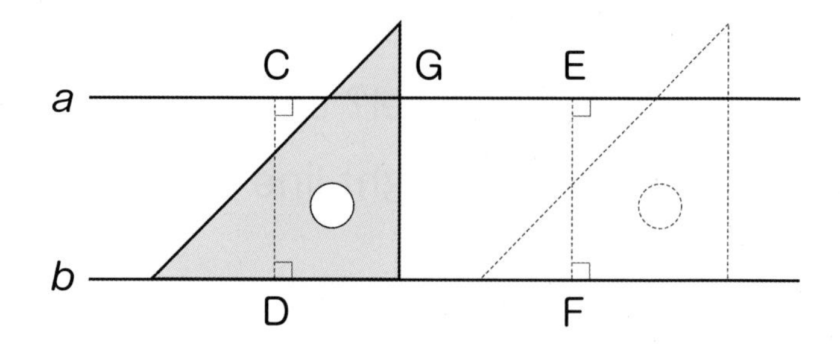

\ Want to explore /

? Purpose **What is the relationship between the two parallel straight lines?**

1. Let's compare the length of lines CD and EF.
2. If you set a triangle ruler on line *b* and slide, how will the place of point G, which intersects with line *a* change?
3. If you extend the straight lines *a* and *b*, will these lines cross?

! Summary

The length between two parallel straight lines is equal at every point. Both straight lines never cross, no matter how far they are extended.

1 Let's explain that the opposite sides, on a rectangle and a square, are parallel.

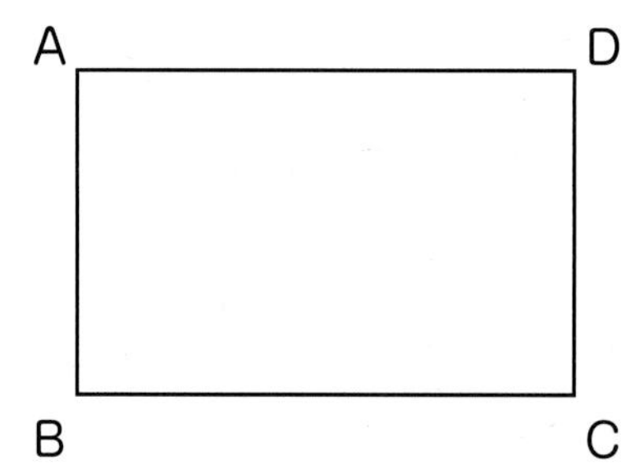

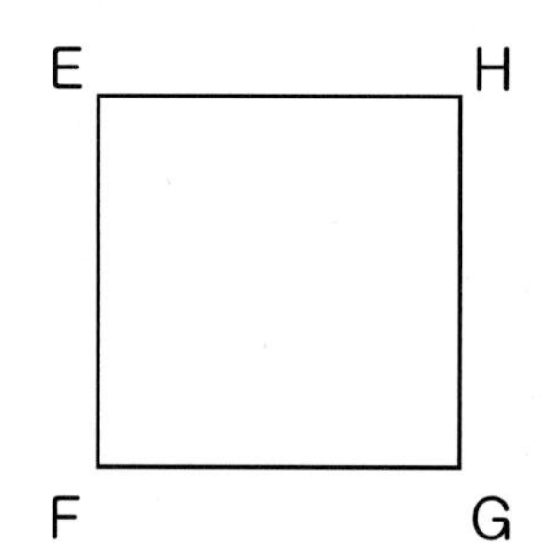

2 Let's look for perpendicular and parallel things from your surroundings.

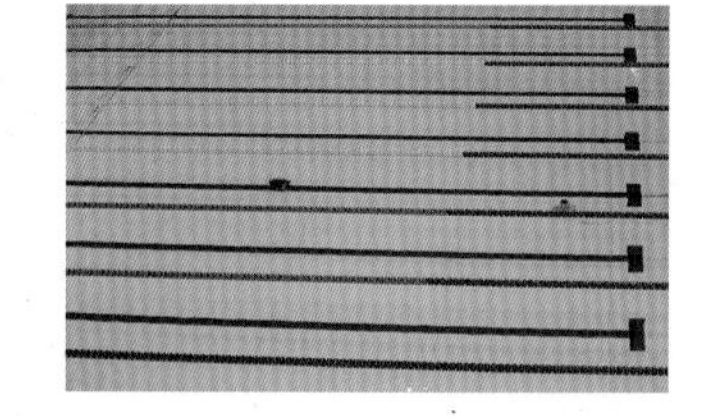

? We could draw a perpendicular line. Can we draw a parallel line too?

3 Let's think about how to draw a parallel straight line to straight line *a*. Let's read the drawing methods of Akari and Yu below and explain the reason why straight lines are parallel in the two cases.

＼ Want to try ／

Purpose Let's draw a parallel line.

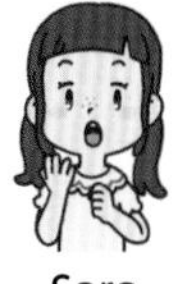

Sara

a ————————————————

What properties do they use?

Akari's way of drawing ▷

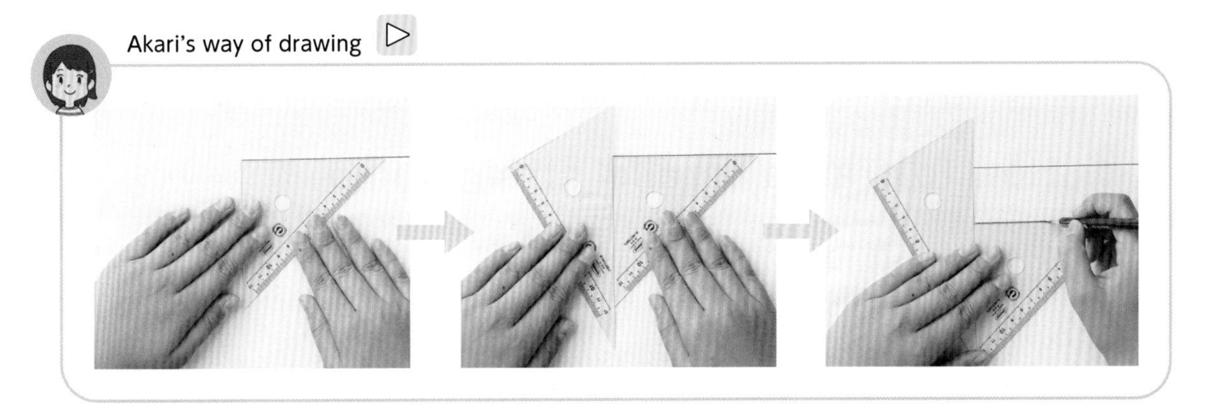

Yu's way of drawing ▷

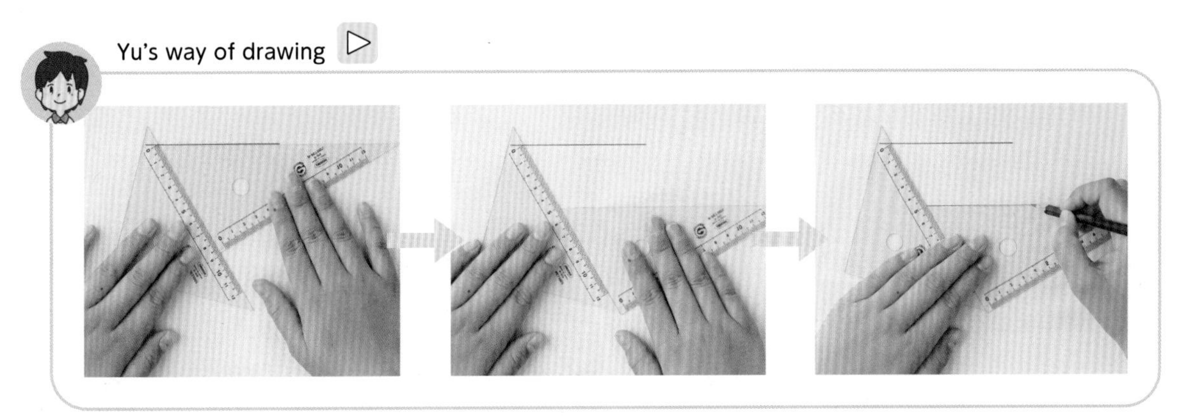

1 Let's draw a parallel straight line to straight line b through point A.

b

•A

How to draw a parallel straight line 

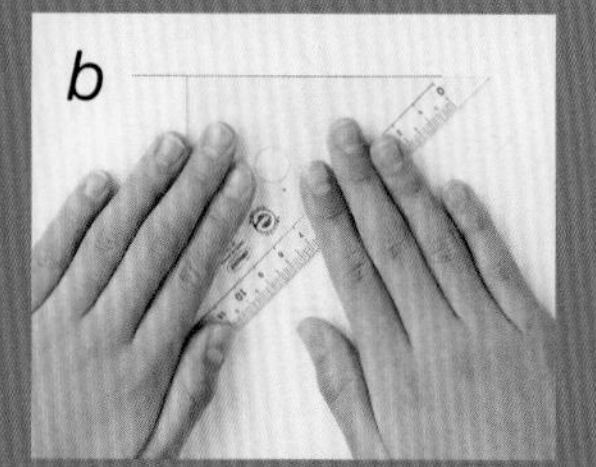

(1) Match the triangle ruler with straight line b.

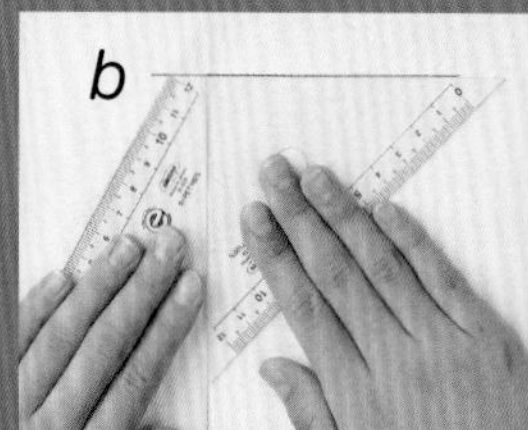

(2) Match the other triangle ruler too.

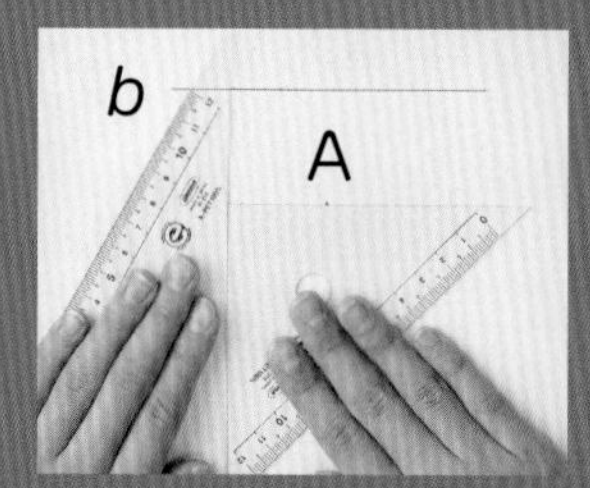

(3) Slide the right hand triangle ruler to match point A.

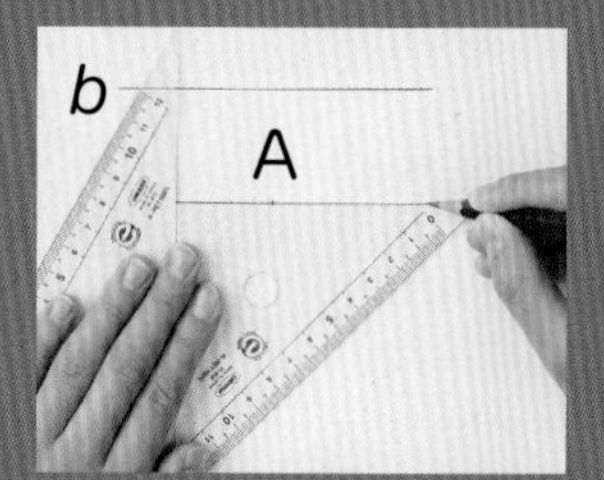

(4) Draw a straight line through point A while holding the triangle ruler.

Summary We can draw parallel lines by using two triangle rulers.

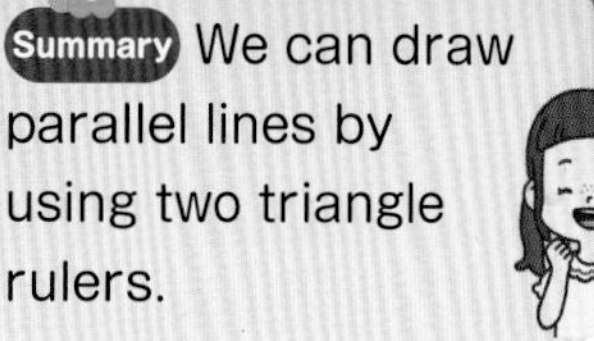

Let's think about why we can draw parallel lines in this way.

2 Let's draw the following straight lines on the diagram below.

① Straight line *d* that goes through point B and is parallel to straight line *c*.

② Straight lines *e* and *f*, each 2cm far from straight line *c* and parallel to it.

c

B

3 Look at the following straight lines. Let's connect the dots to draw parallel straight line for each of them.

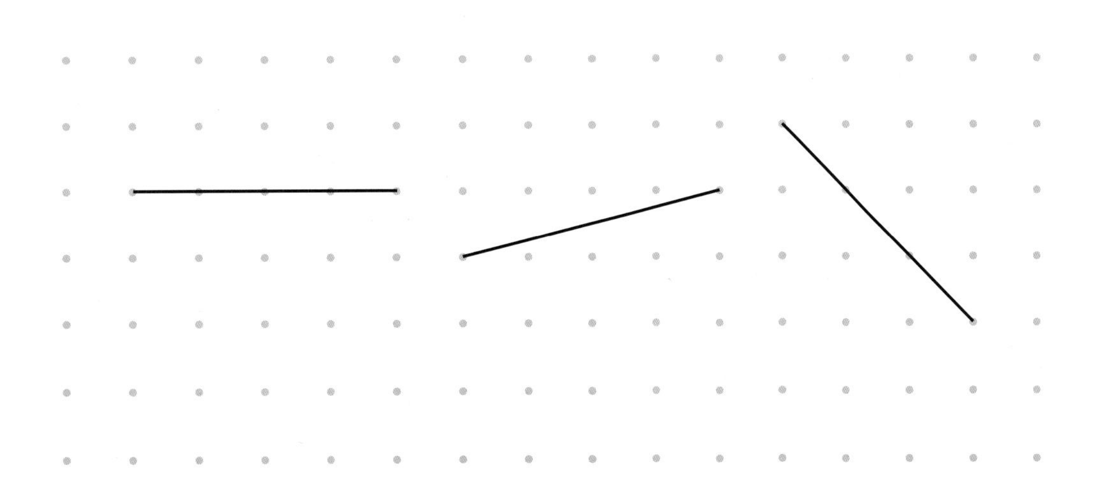

? Can we classify the quadrilaterals based on the perpendicular and parallel lines?

3 Various quadrilaterals

Trapezoid →

1 Let's color the parallel sides of the quadrilaterals on page 113 and classify them.

Ⓑ

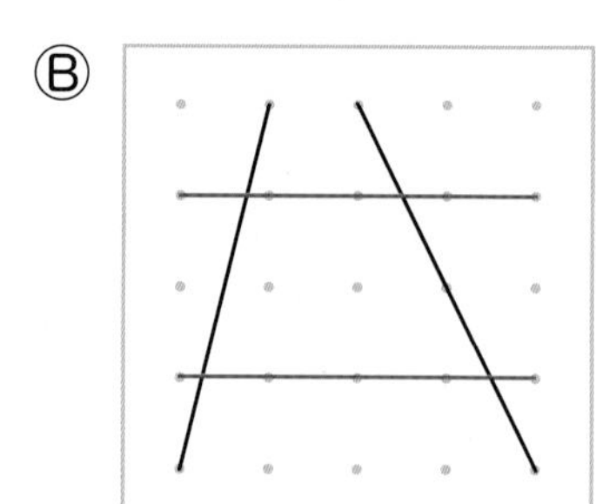

Ⓚ

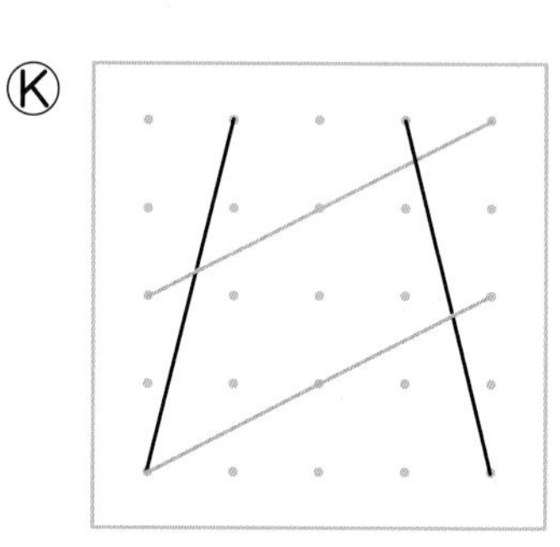

\ Want to know /

Purpose What kind of quadrilaterals have parallel sides?

① Which quadrilaterals on page 113 have one pair of opposite parallel sides?

A quadrilateral that has one pair of opposite parallel sides is called a **trapezoid**.

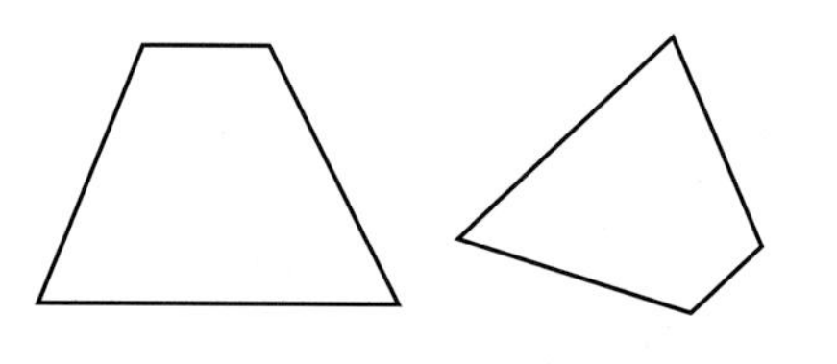

1 Let's draw a trapezoid using two parallel straight lines.

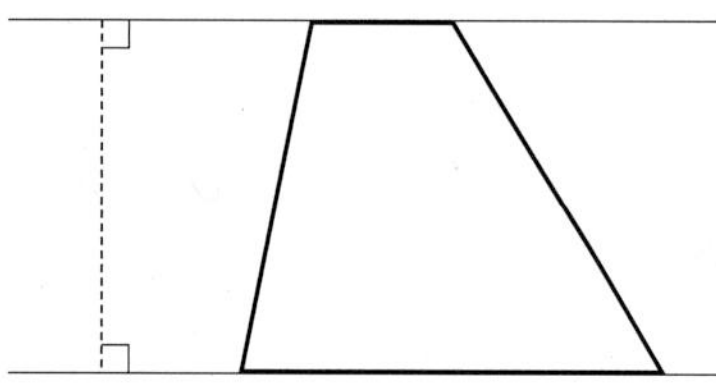

2 Let's look for things with the shape of a trapezoid from your surroundings.

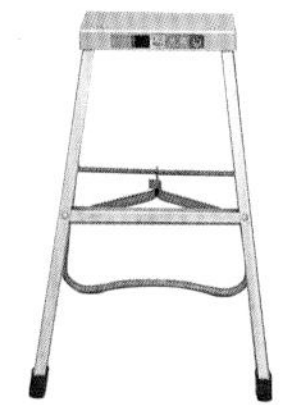

? What is the name of the quadrilateral that has two pairs of opposite parallel sides?

2 Which quadrilaterals on page 113 have two pairs of opposite parallel sides?

Ⓓ
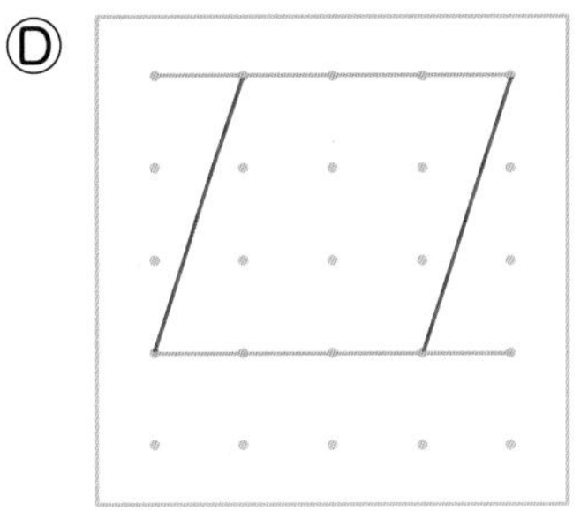

Ⓘ
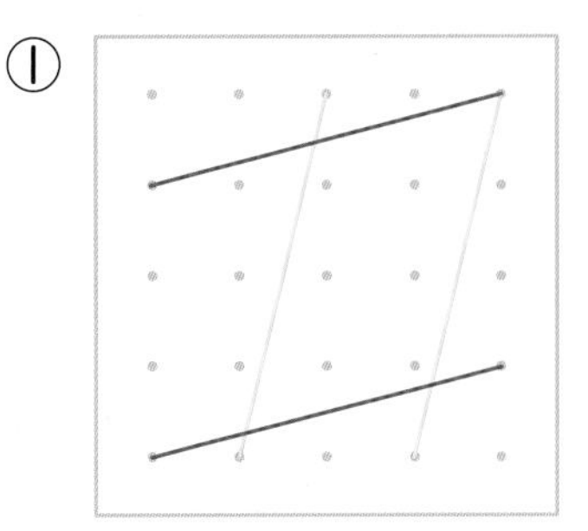

\ Want to know /

Purpose What is the shape of quadrilateral that has two pairs of opposite parallel sides?

Haruto

A quadrilateral that has two pairs of opposite parallel sides is called a **parallelogram**.

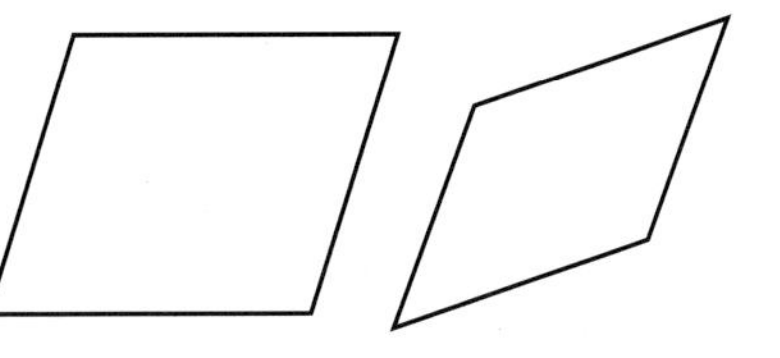

1 Let's draw parallelograms using the following grid paper.

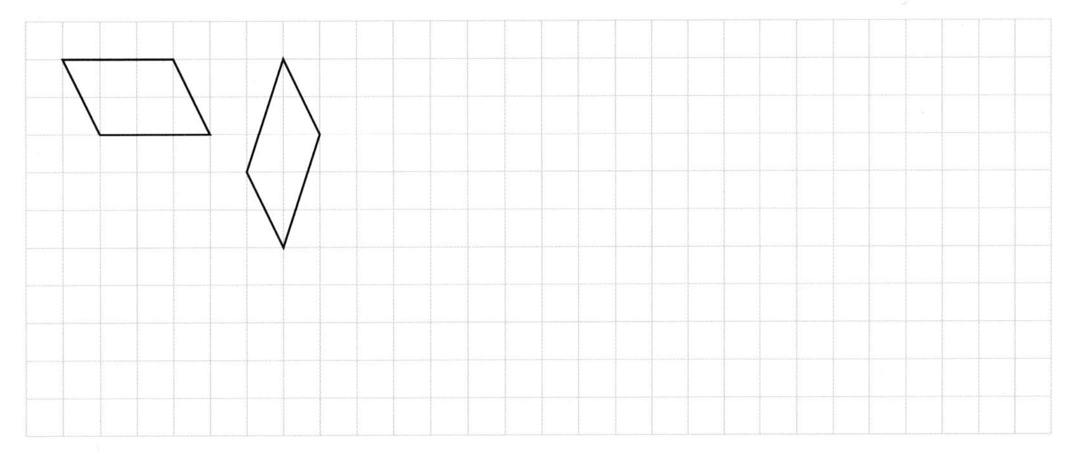

2 Let's look for things with the shape of a parallelogram from your surroundings.

Numazu-gaki
(Numazu City, Shizuoka Pref.)

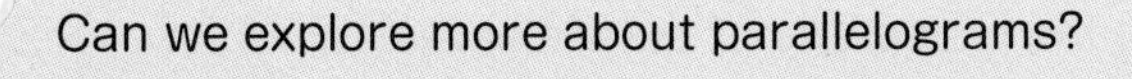

Can we explore more about parallelograms?

\ Want to know /

Purpose What are the properties of a parallelogram?

Sara

3 Let's explore the length of the opposite sides and the size of the opposite angles in a parallelogram.

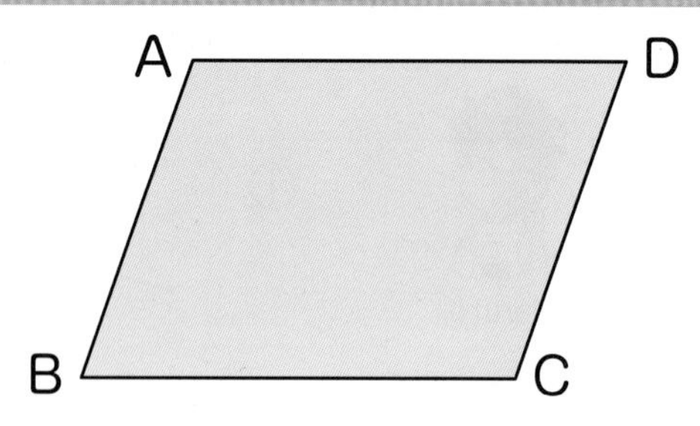

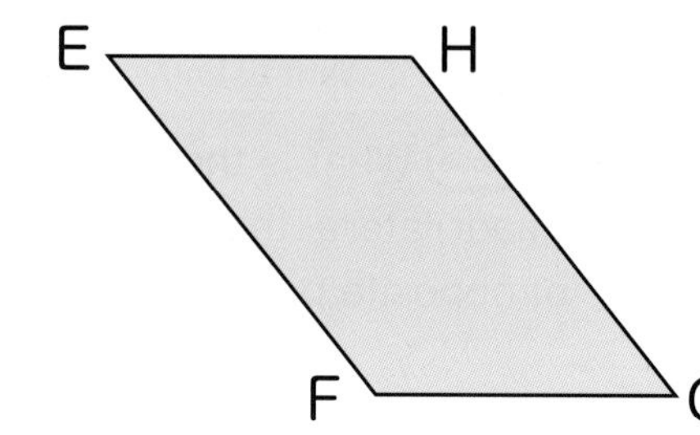

The size of angles A and C looks the same.

Haruto

Akari

We can investigate using a protractor or a compass.

Summary

In parallelograms, the length of the opposite sides are equal and the size of the opposite angles are equal.

1 What is the length (cm) of side AD and side CD shown on the right parallelogram? What is the size of the angles C and D?

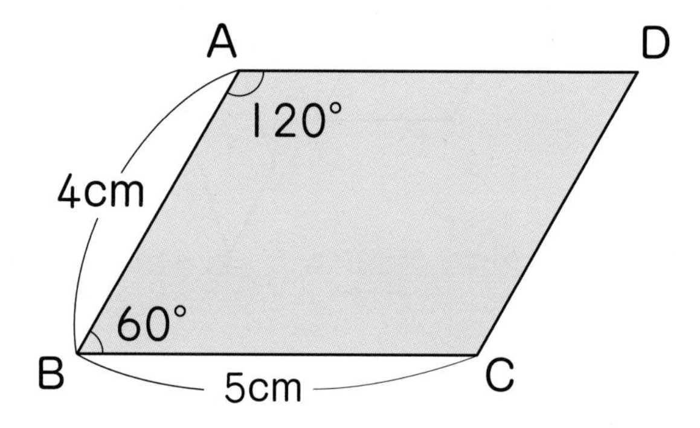

2 Let's find trapezoids and parallelograms from the quadrilaterals on the right.

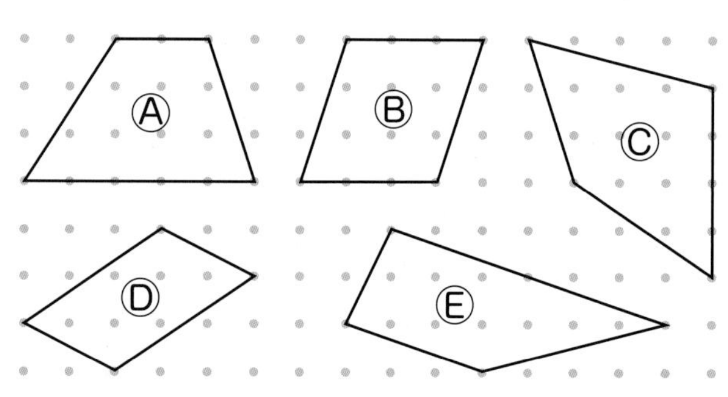

? Can we draw a parallelogram using the properties of parallelograms?

4 Let's think about how to draw a parallelogram like the one shown on the right.

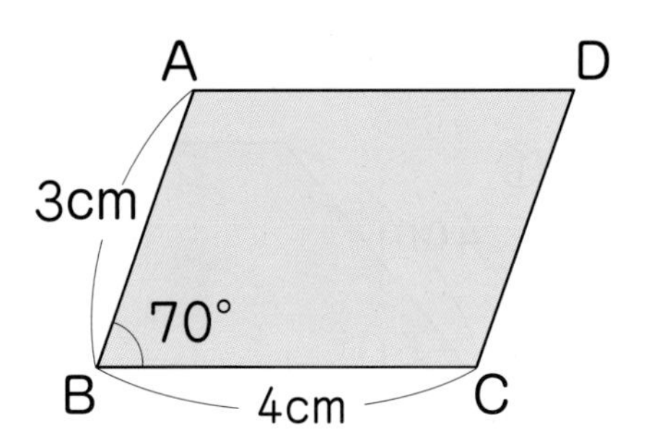

① The vertices A, B, and C are determined by the following drawing method. Let's think about how to determine the position of vertex D.

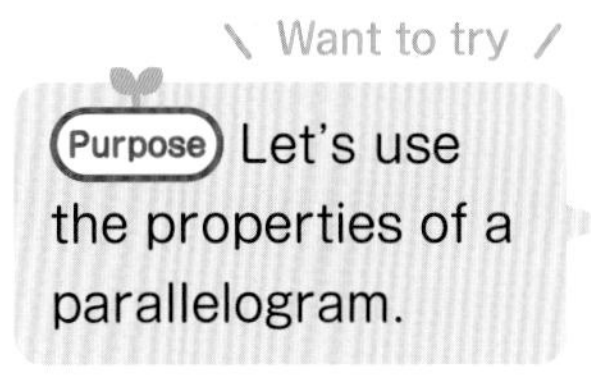

Want to try
Purpose Let's use the properties of a parallelogram.

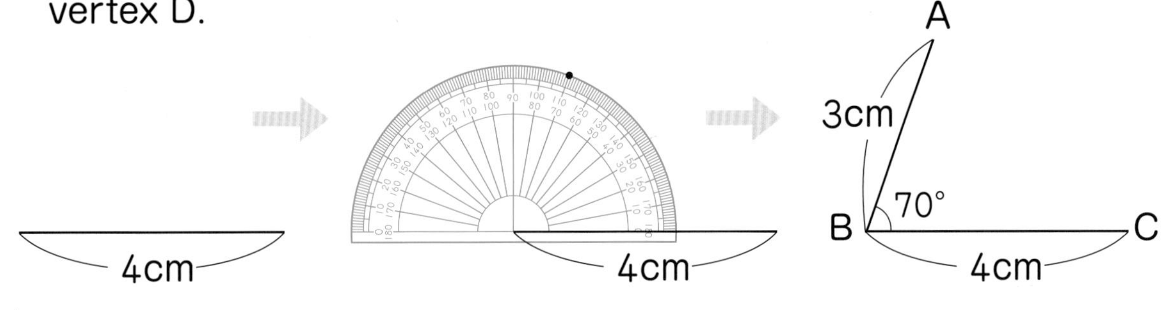

② Let's explain the ideas of Yu, Sara, and Akari.

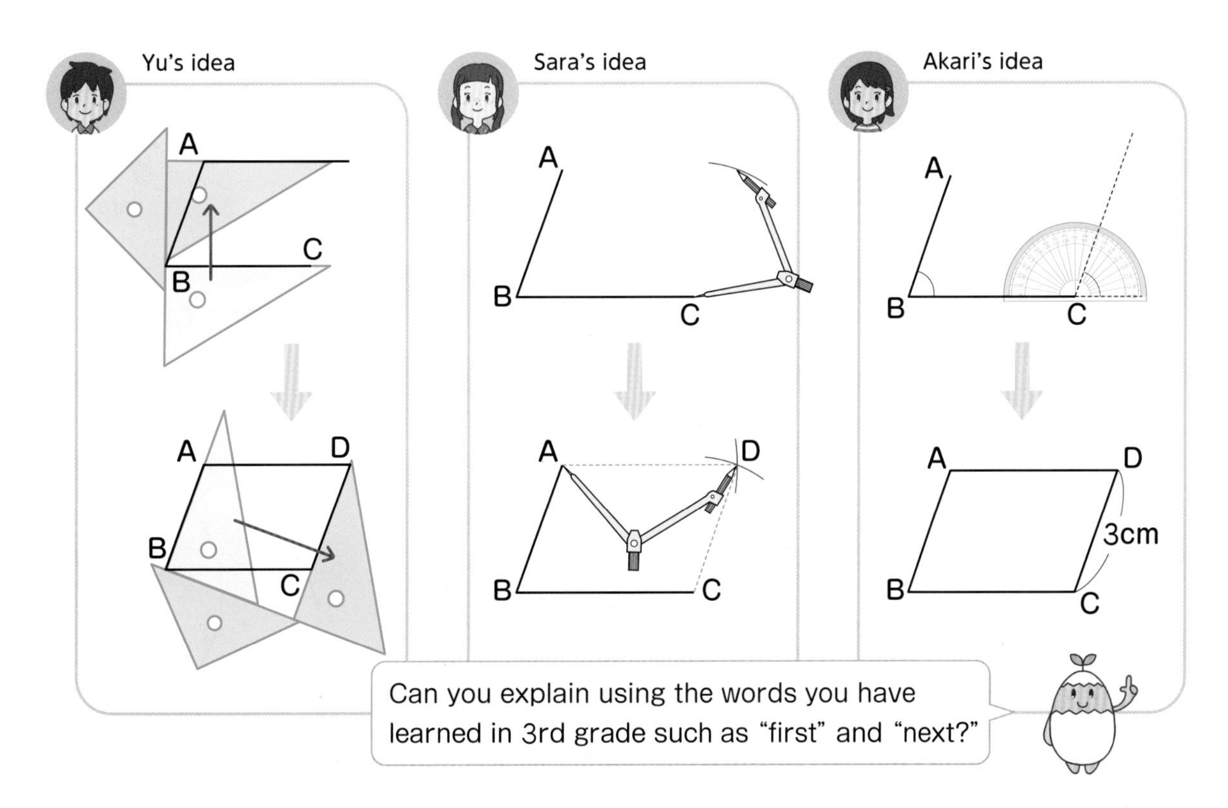

1 Let's draw the following parallelograms.

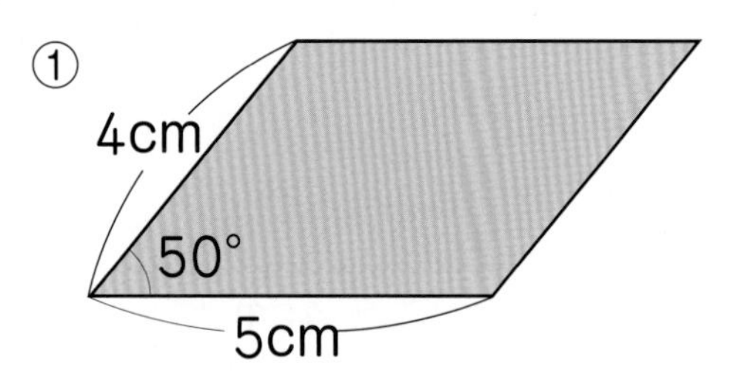

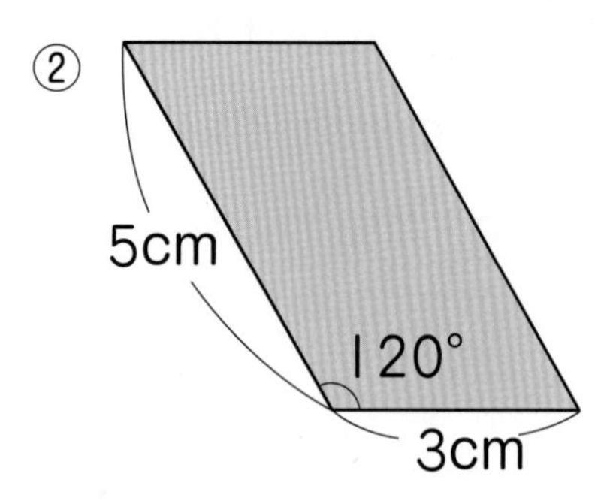

? Is there a name for a quadrilateral with four equal sides?

Rhombus →

5 Let's examine the length of the four sides of quadrilateral Ⓒ and Ⓙ on page 113.

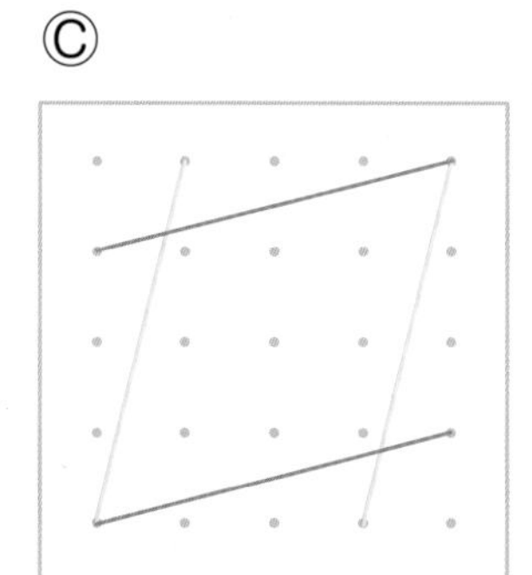

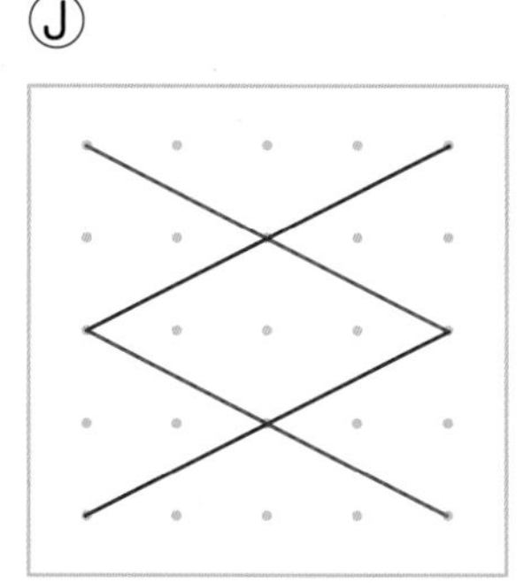

A quadrilateral with four equal sides is called a **rhombus**.

Want to know
Purpose What are the properties of a rhombus?
Sara

1 The diagram below shows two parts of perimeters of circles centered at A and C, and with the same radius. They intersect at B and D. Let's explore this diagram.

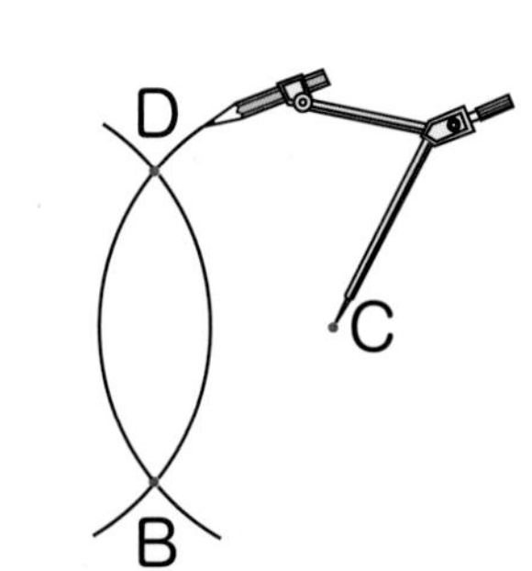

① Let's connect the points A → B → C → D → A with straight lines to draw a quadrilateral.

② Let's examine the lengths of the sides and size of the angles. Which quadrilateral is this?

2 Let's explore the rhombus that you drew on exercise **1** at the bottom of the previous page.

① Are the sizes of opposite angles equal?

② Are the opposite sides parallel?

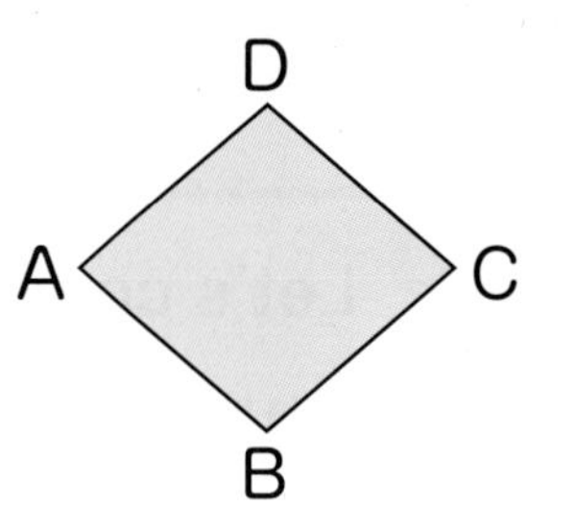

Summary

In a rhombus, the size of opposite angles are equal and the opposite sides are parallel.

3 What is the length (cm) of sides AB, BC, and CD shown on the rhombus on the right? What are the size of the angles A and D?

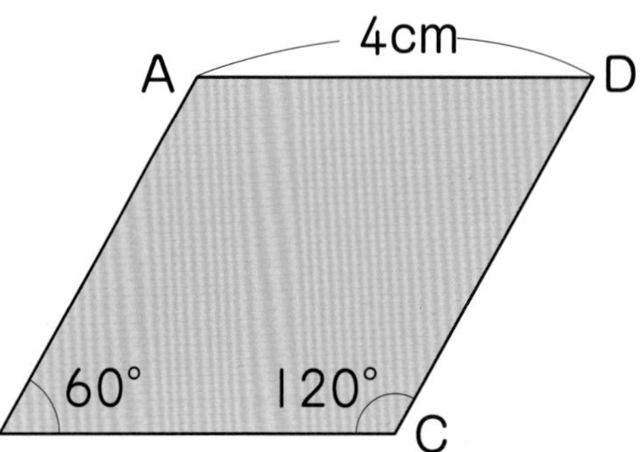

4 Let's think about how to draw the rhombus on the right.

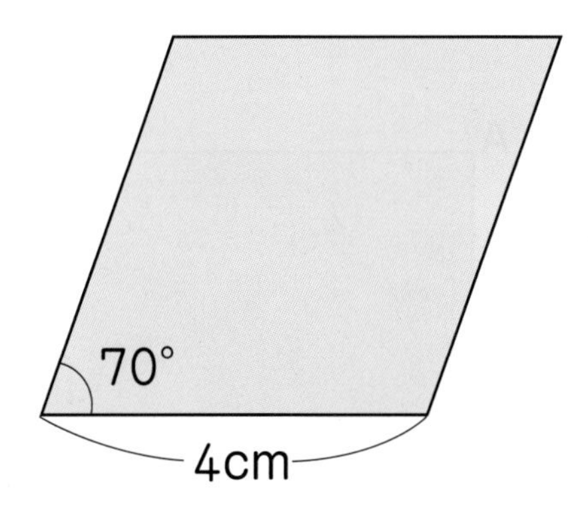

5 Let's look for things with the shape of a rhombus from your surroundings.

koginzashi stitch

? Do various quadrilaterals have other properties?

4 Diagonals of quadrilaterals

Let's connect the opposite vertices of the quadrilaterals with straight lines. Let's discuss what you have noticed.

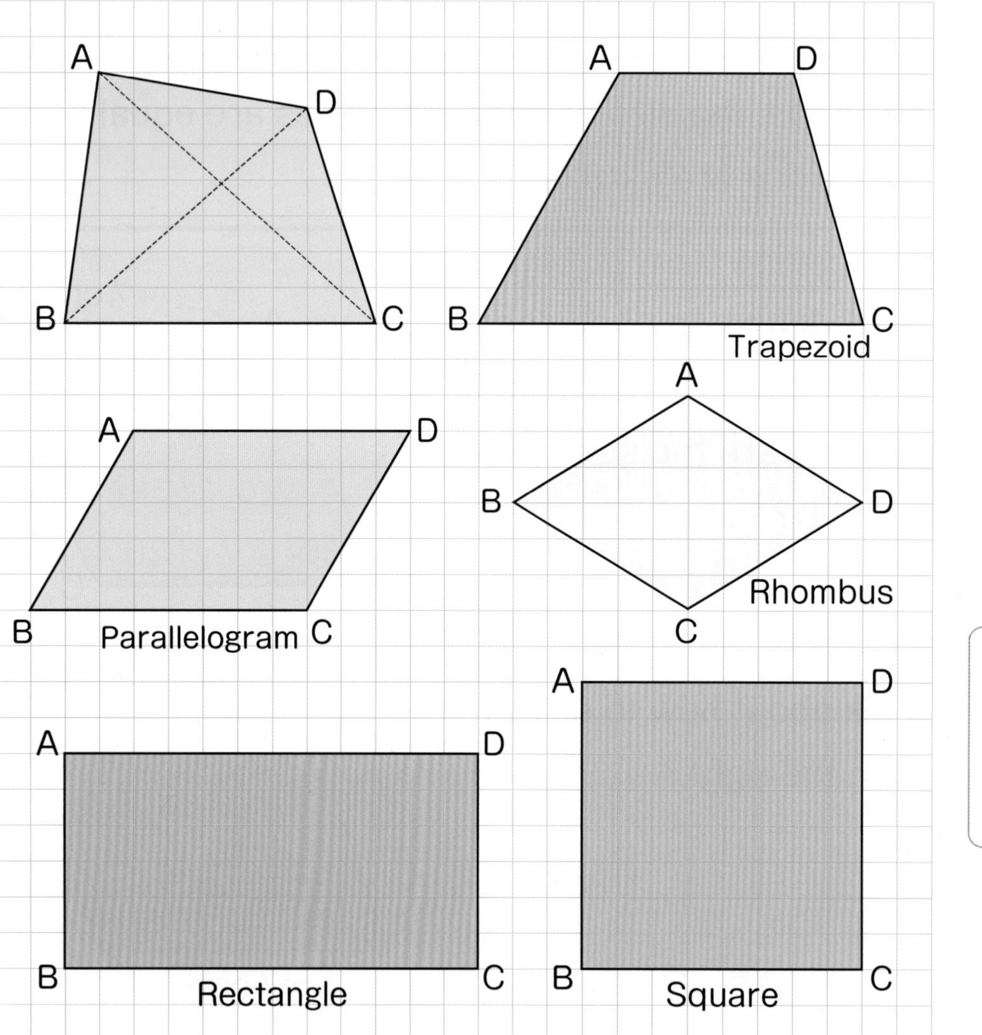

Some shapes have the same properties.

Sara

Each straight line that you draw to connect the opposite vertices is called a **diagonal**. There are two diagonals in each quadrilateral.

\ Want to explore /

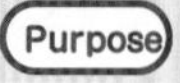
Purpose What is the length and the angles of the diagonal?

Haruto

1 Let's summarize the properties of the quadrilateral's diagonals that you found in exercise **1** on the previous page in the table below. Write a ○ on what always applies.

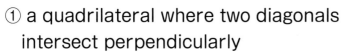

Name of the quadrilateral / Properties of the quadrilateral's diagonals	Trapezoid	Parallelogram	Rhombus	Rectangle	Square
① Both diagonals have equal length.					○
② Both dianonals intersect at their midpoint.					○
③ Both diagonals are perpendicular lines.					○

2 Draw the following quadrilaterals using the properties investigated in **1**.

① a rhombus with 4cm and 3cm length diagonals

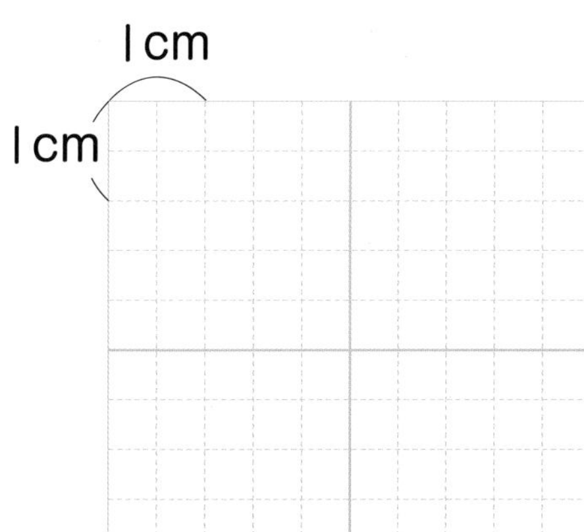

② a square with 4cm length diagonals

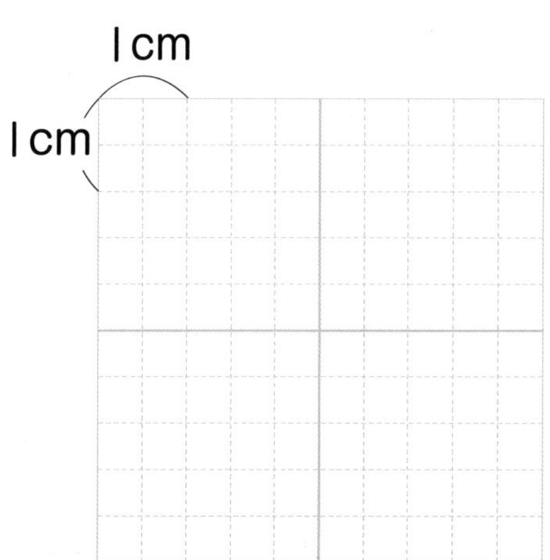

? Some quadrilaterals have similarities. What are the relationship?

5 Relationship of quadrilaterals

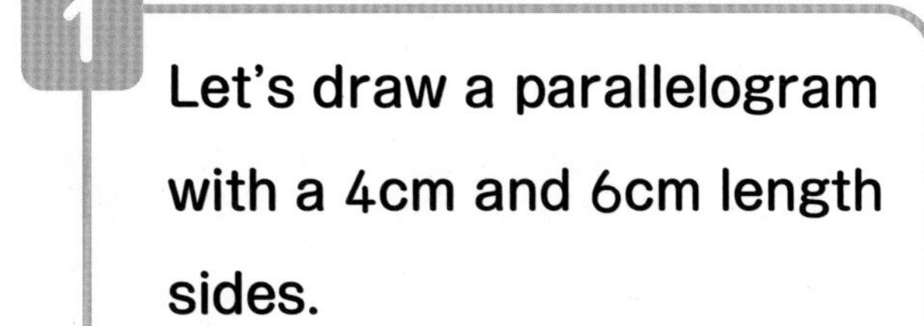

1 Let's draw a parallelogram with a 4cm and 6cm length sides.

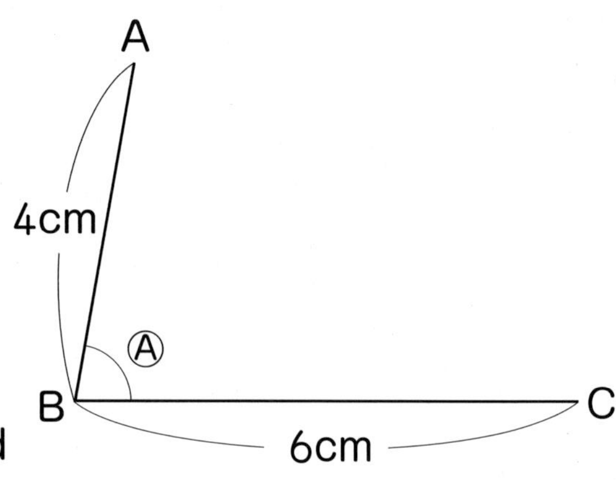

❶ Let's draw angle Ⓐ with 80° and 120°.

❷ Let's draw angle Ⓐ with 90°.

What kind of quadrilateral can be drawn?

1 Let's draw a rhombus with a 5cm length side.

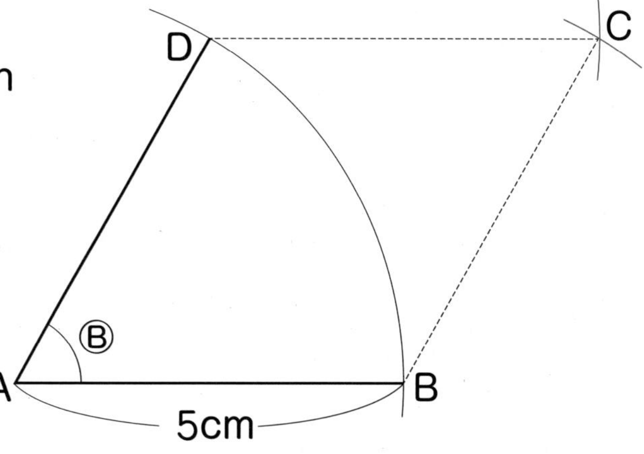

① Let's draw angle Ⓑ with 60°.

② Let's draw angle Ⓑ with 120°.

③ Let's draw angle Ⓑ with 90°.

What kind of quadrilateral can be drawn?

What are the sizes of the 4 angles in each drawing?

\ Want to think /

Purpose What can we know from the size of the angles and the length of the sides of various quadrilaterals?

Yu

2 Let's discuss what you have learned so far about quadrilaterals, based on the properties you found in page 131 and what you have noticed from page 132. Let's talk about the similarities and differences of various quadrilaterals.

Parallelograms and rectangles have two diagonals that intersect at their midpoint.
The diagonals of squares and rhombuses are perpendicular lines.
Squares and rectangles have the same properties on both diagonals.
How about comparing trapezoids with other quadrilaterals?

That's it! Relationship of quadrilaterals

ADVANCED Junior High School

The following classification of various quadrilaterals learned so far can be made based on the length of sides and the size of the angles.

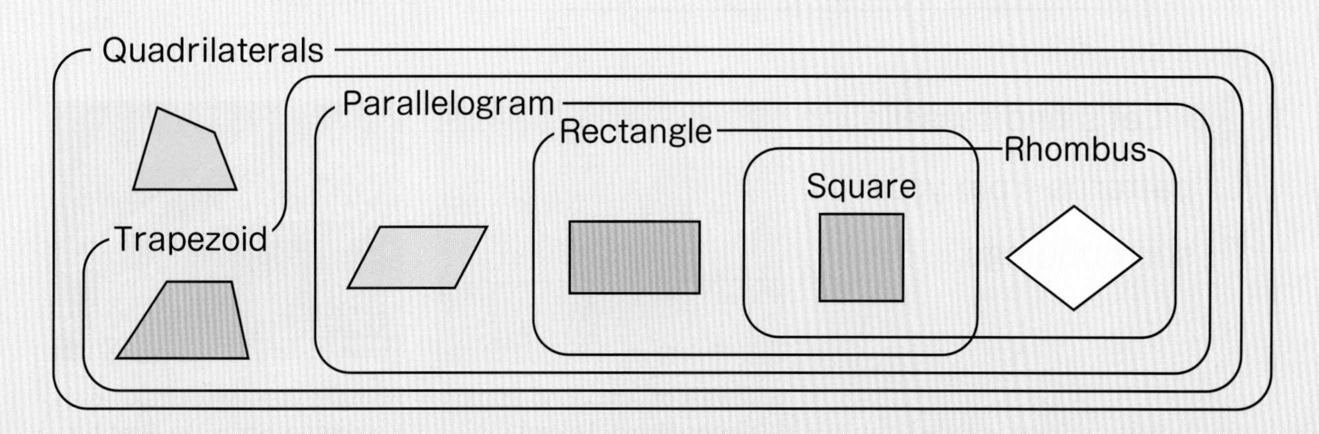

6 Patterns by tessellation

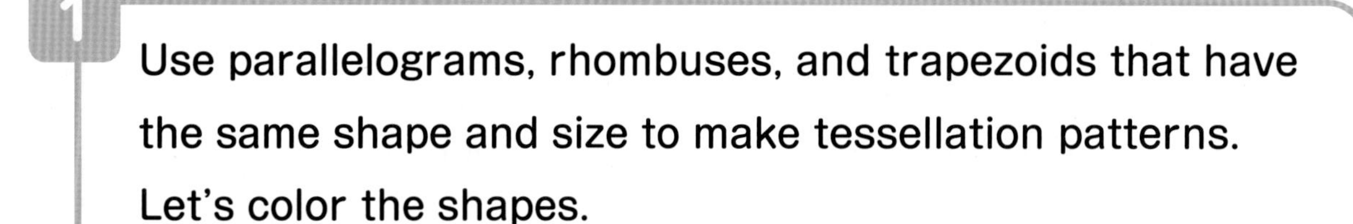

1 Use parallelograms, rhombuses, and trapezoids that have the same shape and size to make tessellation patterns. Let's color the shapes.

①

② 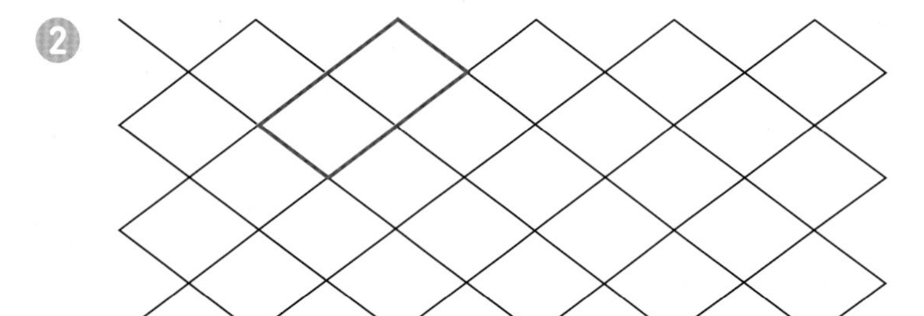

③ 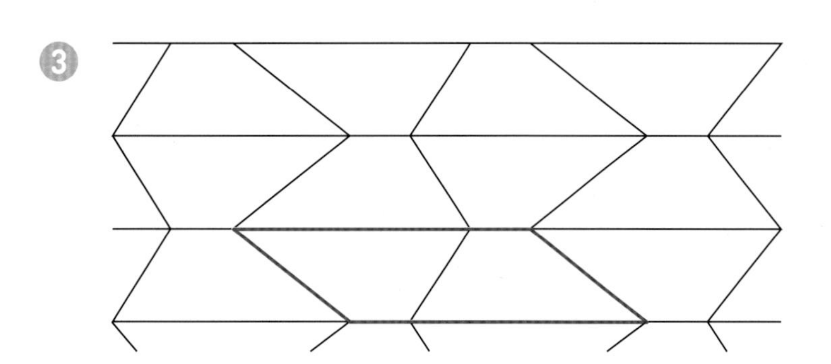

\ Want to explore /

Purpose Let's find out various shapes from the pattern.

Akari

Putting two parallelograms together forms a parallelogram?

Yu

I found a parallelogram inside the rhombus pattern.

Sara

I also found a parallelogram inside the trapezoid pattrn.

Haruto

Why do all of them become a parallelogram?

Akari

1 Let's find tessellation patterns from your surroundings.

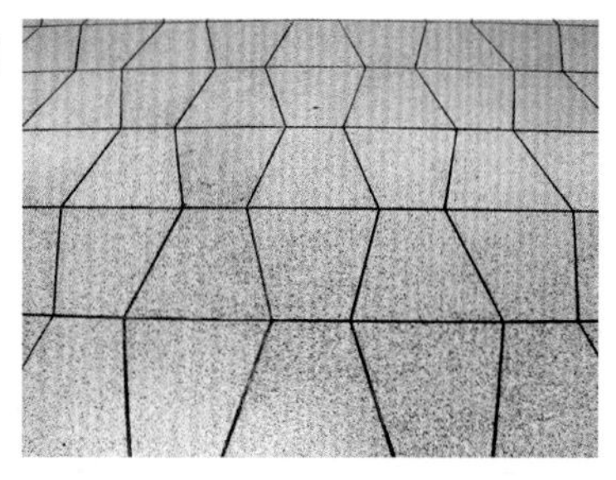

Sidewalk of Jimbocho Station (Chiyoda Ward, Tokyo)

Namako Wall (Matsuzaki Town, Kamo-gun, Shizuoka Pref.)

CAN What can you do?

☐ We understand how to draw perpendicular and parallel straight lines. → p.117, 122

1 Let's draw perpendicular and parallel straight lines passing through point A to straight line *a*.

A

a

☐ We understand the properties of various quadrilaterals. → p.124, 125, 128, 131

2 Let's answer the following questions.

① Look at the figures on the right, and fill in the ☐ with appropriate words.

ⓐ A quadrilateral that has one pair of ☐ opposite sides is called a ☐.

ⓑ A quadrilateral that has two pairs of ☐ opposite sides is called a ☐.

ⓒ A quadrilateral that has four sides with ☐ length is called ☐.

ⓐ ⓑ ⓒ

② Let's draw a rhombus with 5cm and 3cm length diagonals.

1cm 1cm

Supplementary Problems → p.149

Which "Way to See and Think Monsters" did you find in "9 Perpendicular, Parallel and Quadrilaterals"?

I found "Summarize" when I was trying to investigate various quadrilaterals.

Yu

I found other monsters too!

Utilize

Usefulness and Efficiency of Learning

1 Let's identify two perpendicular lines and the parallel lines in the figure on the right.

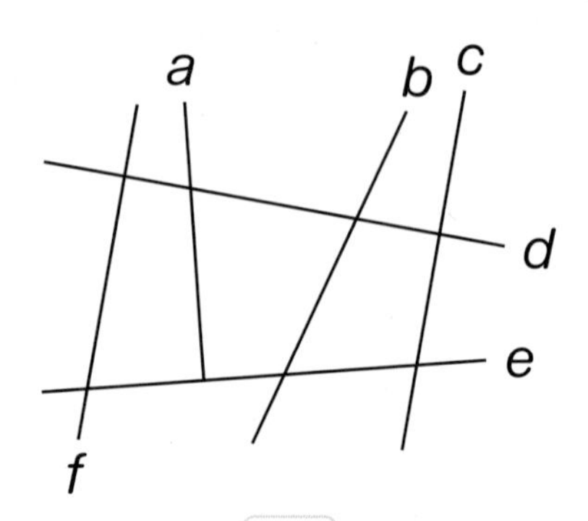

2 There is a parallelogram as shown on the right. Let's fill in the ☐ with numbers.
Let's draw the same parallelogram as shown on the right.

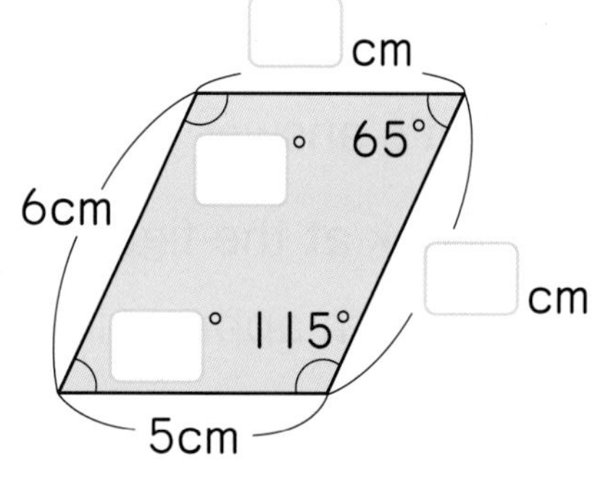

3 Which of these quadrilaterals have the following properties?

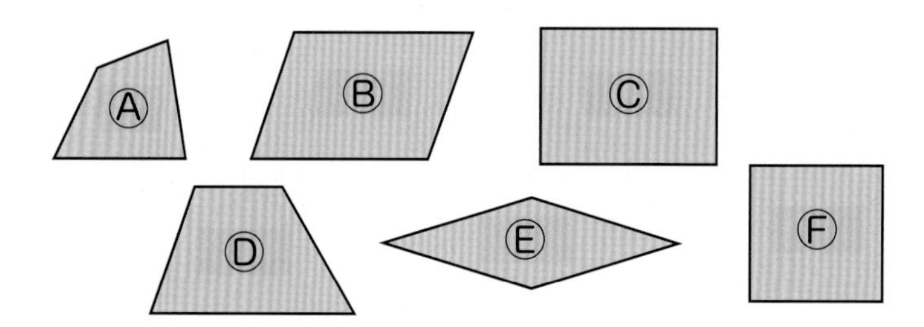

① two pairs of parallel sides
② four angles with equal size
③ diagonals with equal length
④ opposite sides with equal length
⑤ opposite angles with equal size
⑥ no parallel sides

4 The following shows the diagonals of various quadrilaterals. Let's identify the names of the quadrilaterals by measuring the length and the size of angles.

①

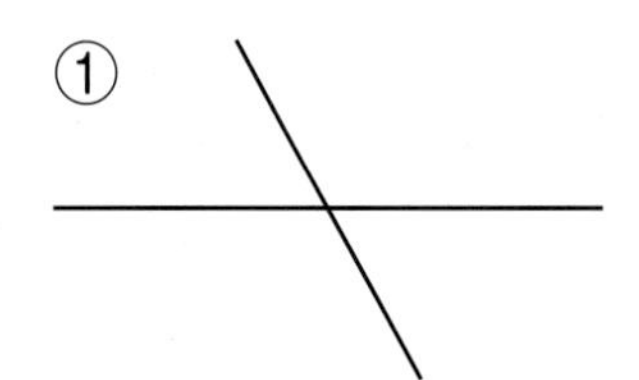

②

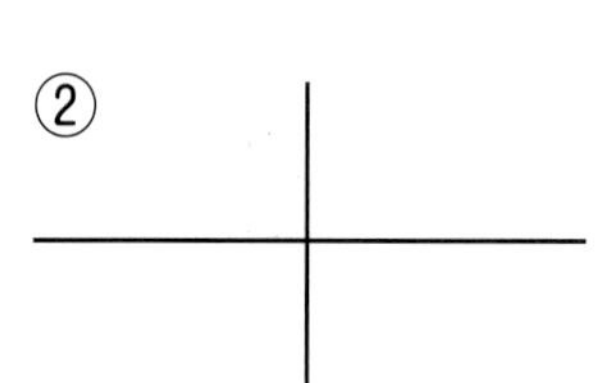

③

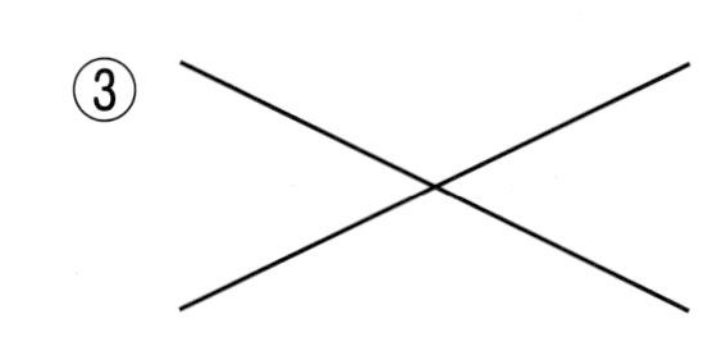

With the Way to See and Think Monsters...

Let's Reflect!

Let's reflect on which monster you used while learning "9 Perpendicular, Parallel, and Quadrilaterals."

Summarize

We could summarize quadrilaterals into trapezoids, parallelograms, and rhombuses, depending on their properties such as the length of the sides and whether the two opposite sides are parallel or not.

Can we say that parallelograms and rhombuses are in the same group?

① What kind of quadrilaterals can trapezoids, parallelograms, and rhombuses be explained as?

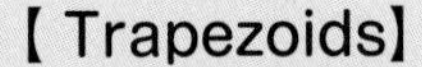

【Trapezoids】

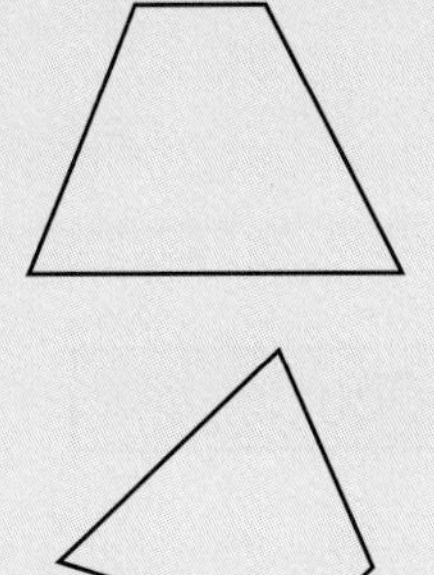

【Parallelograms】

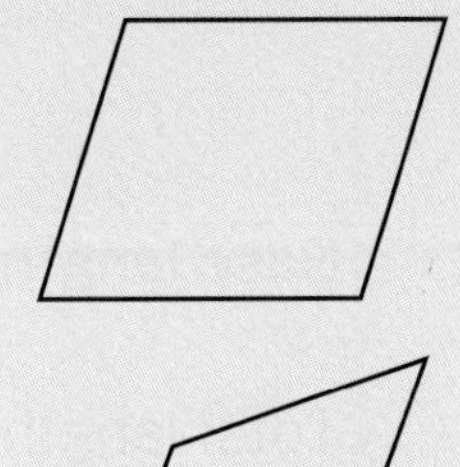

【Rhombuses】

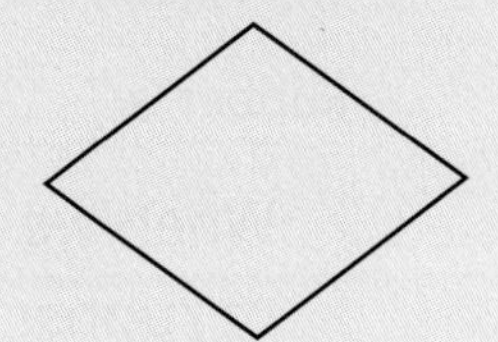

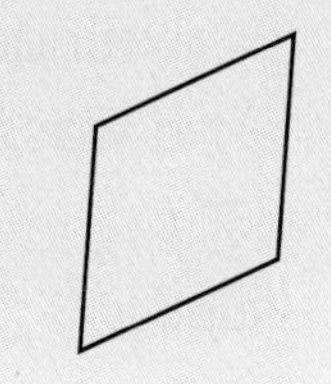

Yu: These are quadrilaterals that one pair of opposite parallel sides.

Akari: These are quadrilaterals that has two pairs of opposite parallel sides.

Haruto: The length of the opposite sides are equal, and the size of opposite angles are equal.

Sara: These are quadrilaterals that four length of the sides are equal.

Let's deepen. → p.153

Solve the ?

We could identify the quadrilaterals by investigating whether the sides are perpendicular or parallel, and the properties of the diagonals.

Akari

→

Want to Connect

Can we investigate other shapes?

Sara

Let's Try!

How many times (2) – Introduction to Ratio –

Let's think about how many times by comparing the length of rubber.

1 There are two rubbers ⓐ and ⓑ. We want to compare which rubber extends more. Let's answer the following questions.

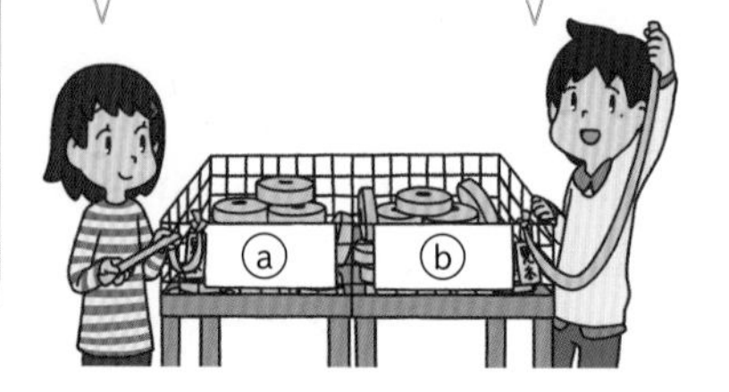

I want a rubber that extends a lot.
There are two types of rubbers.
ⓐ
ⓑ

❶ We extended rubbers ⓐ and ⓑ as follows. Which rubber extends more?

Rubber ⓐ

When we cut 50cm of rubber ⓐ and extended, it was 150cm.

Base length: 50cm
Total length: 150cm

Rubber ⓑ

When we cut 100cm of rubber ⓑ and extended, it was 200cm.

Base length: 100cm
Total length: 200cm

As for rubber ⓐ, $150-50=100$, so the extension length is 100cm. As for rubber ⓑ, $200-100=100$, so the extension length is 100cm. — Haruto

The extension length is the same, but the base length is different. — Akari

It would be easier if we can use the same base length compare. — Yu

Want to compare

? **Purpose** How can we compare which rubber extended more?

❷ If we consider the base length of rubbers ⓐ and ⓑ as 1, how can we represent the length after the extension?

Rubber ⓐ

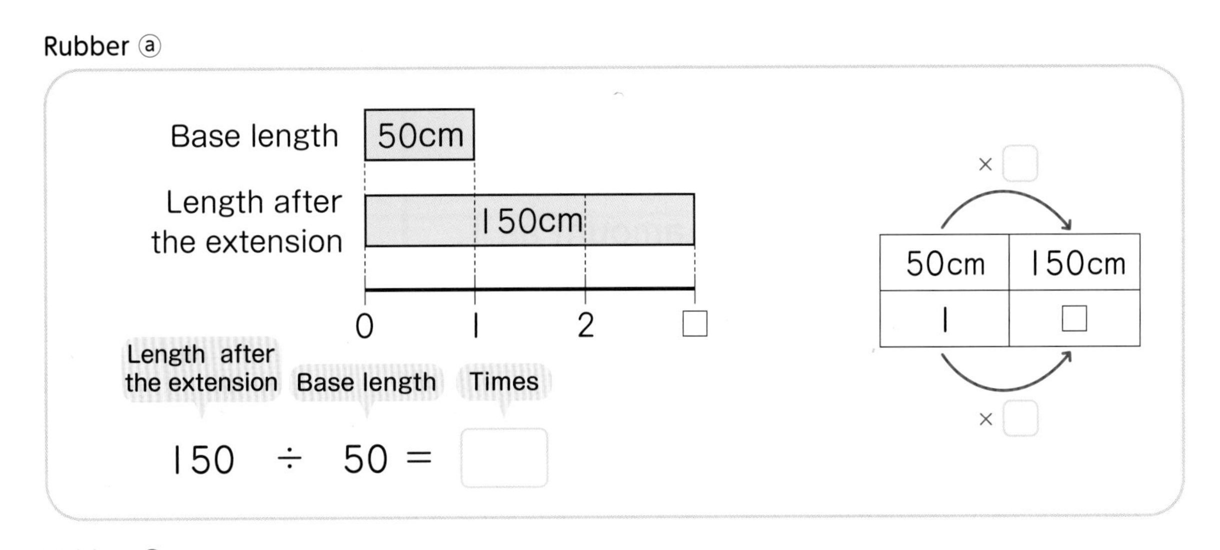

Rubber ⓑ

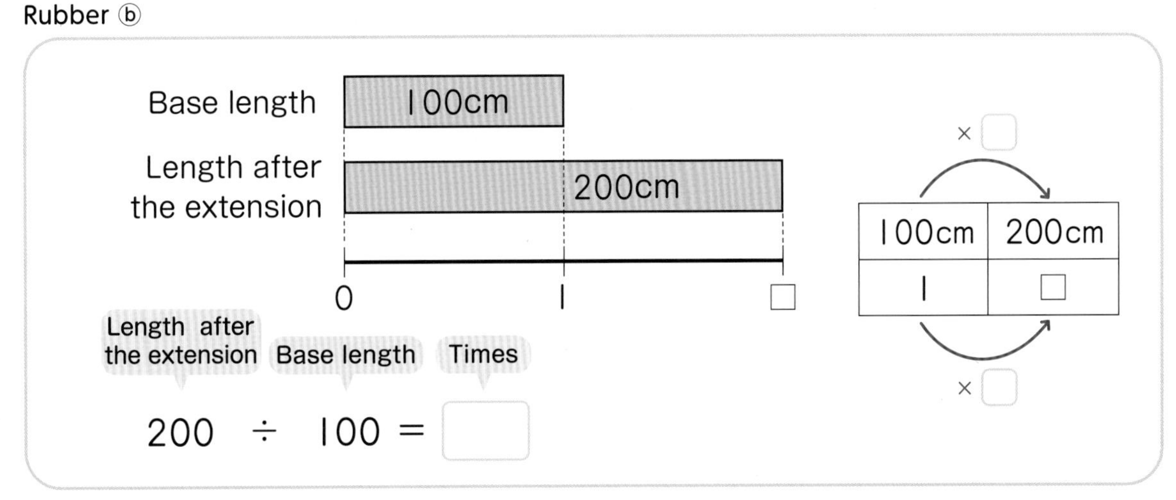

When a certain quantity is represented by how many times of the base quantity, "times of" is also called **ratio**.

③ Based on the discussions so far, let's discuss which rubber extends more, ⓐ or ⓑ?

Way to see and think

When you compare, the conclusion depends on whether you compare the difference or "times of".

Summary

If the base amount is different, we can use ratio to compare by considering the base amount as 1.

We can compare the difference by cutting 100cm of rubber ⓐ and extending it.
Akari

1 At a supermarket, the price of bananas increased from 120 yen to 240 yen, and that of the apples from 60 yen to 180 yen. Which fruit became more expensive?

The original price is different, so...
Sara

2 Let's look for things that are compared by difference or ratio from your surroundings.

Haruto
When I get more pocket money, I tend to think in terms of difference.

I've heard on the news that the amount of pollen doubled from the previous year.
Akari

More Math!

[Supplementary Problems]

[Let's deepen.]

[Answers]

1 Large Numbers

→ pp.12 ~ 24

1 Let's spell out by the numeral for the following numbers.

① the number that gathers 305 sets of 100 million

② the sum of 47 sets of 1 billion and 6900 sets of 10 thousand

③ the number that gathers 321 sets of 1 trillion

④ the sum of 5 sets of 10 trillion and 980 sets of 100 million

2 Let's fill in each ☐ with a number.

① 38 billion gathers ☐ sets of 1 billion.

② The number that is 1 million smaller than 100 million is ☐.

③ 29 trillion gathers ☐ sets of 1 trillion.

④ 1 trillion gathers ☐ sets of 10 billion.

3 Let's write the following numbers.

① 10 times 30 thousand

② 100 times 5 billion

③ $\frac{1}{10}$ of 40 billion

④ 10 times 7 billion

⑤ 10 times 80 million

⑥ $\frac{1}{10}$ of 5 trillion

4 Let's fill in each ☐ with appropriate inequality signs.

① 5163800 ☐ 47291000

② 7243000 ☐ 7432000

③ 38725000 ☐ 38270000

④ 164309000 ☐ 164306000

5 Let's make the largest and the smallest whole numbers using all the five numerals: 0, 1, 2, 3, 3.

6 Let's calculate the following.

① 3 million 850 thousand + 2 million 560 thousand

② 26 billion 400 million − 19 billion 700 million

③ 45 trillion × 10

④ 68 billion ÷ 10

2 Line Graphs

→ pp.25 ~ 35

1 Look at the line graph on the right and let's answer the following questions about the temperature changes in a day.

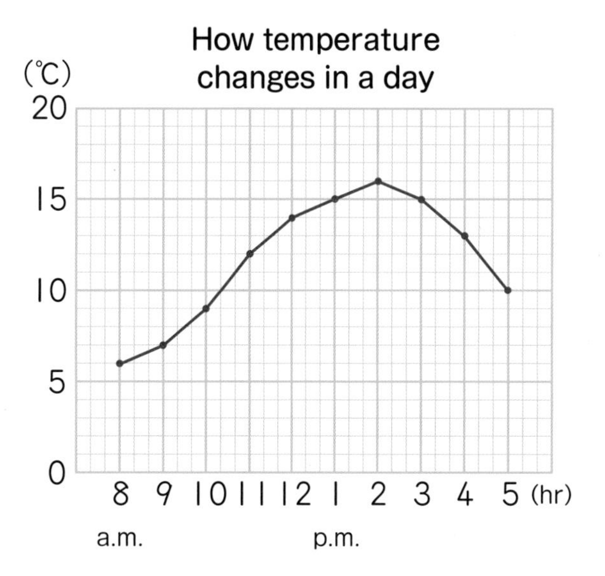

① What was the temperature in ℃ at 10 a.m.?

② What time was the temperature 15℃ ?

③ What was the highest temperature in ℃ and what time?

④ From what time to what time did the temperature rise the most?

⑤ By how many ℃ did the temperature fall between 4 p.m. and 5 p.m.?

⑥ What is the difference in temperature in ℃ between 9 a.m. and 1 p.m.?

2 Considering the situations of Ⓐ~Ⓓ, which is good to be represented by a line graph?

Ⓐ Favorite fruits of my classmates

Ⓑ Body temperature taken every two hours when you have a cold

Ⓒ The temperature of 6 places in school at 10 a.m.

Ⓓ Body weight per month

3 The following table shows how the temperature changes in one year in Shizuoka City, Shizuoka Prefecture. Let's draw a line graph.

How temperature changes in Shizuoka City in one year

Month	1	2	3	4	5	6	7	8	9	10	11	12
Temperature (℃)	7	8	9	15	20	22	28	28	24	19	14	8

How temperature changes in Shizuoka City in one year

(℃)
30
25
20
15
10
5
0
1 2 3 4 5 6 7 8 9 10 11 12
(month)

3 Division

→ pp.38 ～ 45

1 Let's fill in each □ with a number.

① $36 \div 4 = 9$, ↓ $\div$ ⓐ, ↓ $\div$ ⓑ, $9 \div 1 =$ ⓒ

② $24 \div 8 = 3$, ↓ $\div$ ⓐ, ↓ $\times$ ⓑ, $24 \div 4 =$ ⓒ

③ $6 \div 2 = 3$, ↓ $\times$ ⓐ, ↓ $\times$ ⓑ, $18 \div 6 =$ ⓒ

④ $8 \div 4 = 2$, ↓ $\times$ ⓐ, ↓ $\div$ ⓑ, $8 \div 8 =$ ⓒ

⑤ $42 \div 7 = 6$, ↓ $\div$ ⓐ, ↓ $\div$ ⓑ, $7 \div 7 =$ ⓒ

⑥ $9 \div 3 = 3$, ↓ $\times$ ⓐ, ↓ $\times$ ⓑ, $27 \div 3 =$ ⓒ

2 Let's fill in each □ with a number.

① $3000 \div 500 = \square \div 5$
$= \square$

② $2800 \div 700 = 28 \div \square$
$= \square$

③ $125 \div 25 = \square \div 50$
$= \square$

3 Let's fill in each □ with a number.

① $80 \div 40$
$= 8 \div \square$
$= \square$

② $90 \div 30$
$= \square \div 3$
$= \square$

③ $160 \div 20$
$= \square \div 2$
$= \square$

4 Angles

→ pp.46 ～ 61

1 Let's fill in each ☐ with a number.

① 1 right angle = ☐ °

② ☐ right angles = 180°

③ 4 right angles = ☐ °

2 What are the sizes of the following angles in degrees?

①

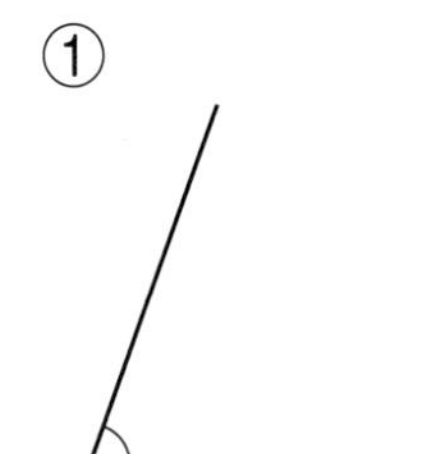

②

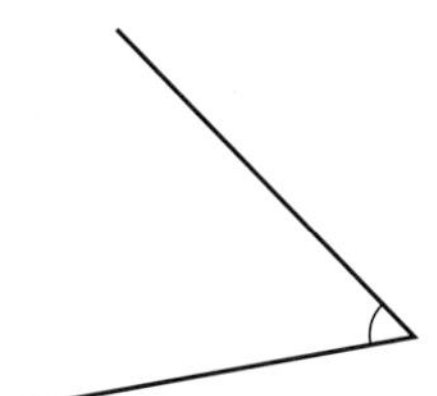

③

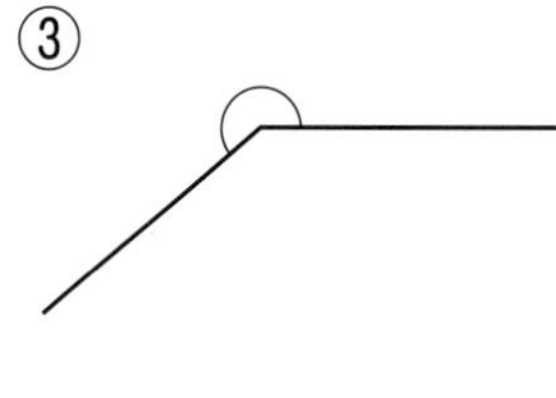

④

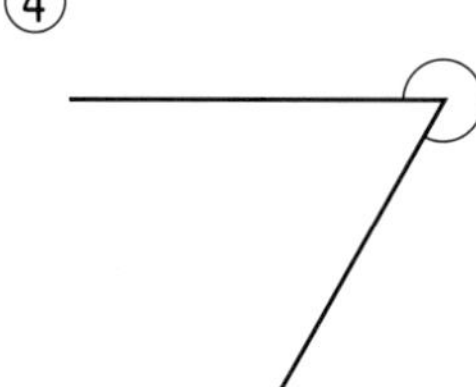

3 Two straight lines intersect as shown on the right. What are the sizes of the angles ⓐ, ⓑ, and ⓒ in degrees?

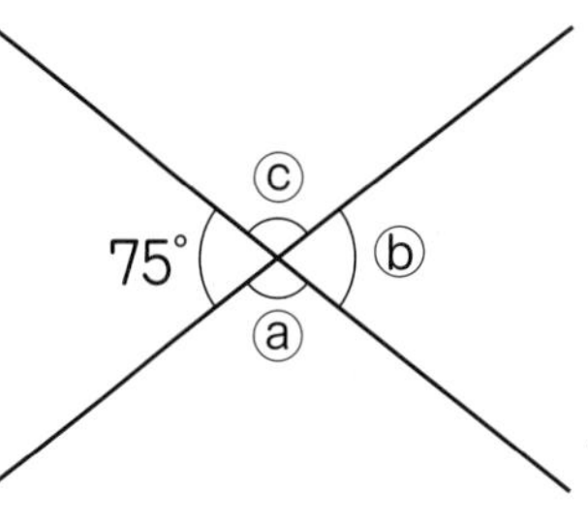

4 Let's draw angles with the following degrees.

① 65°　　② 200°　　③ 320°

5 As shown below, two triangle rulers are combined to make angles. What are the sizes of the angles ⓐ, ⓑ, and ⓒ in degrees?

①

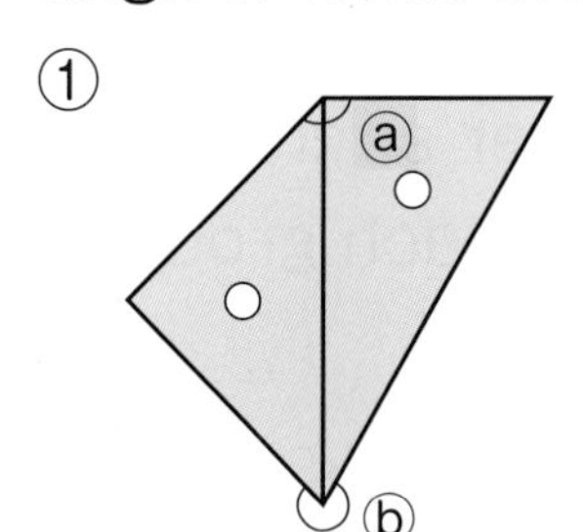

②

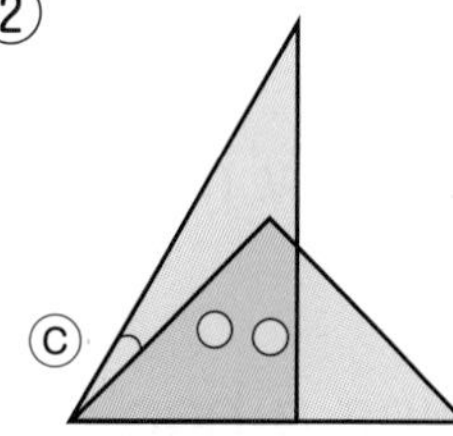

6 Division by 1-digit Numbers

→ pp.66 ~ 78

1 Let's calculate the following in vertical form.

① 62 ÷ 7	② 32 ÷ 5	③ 57 ÷ 8	④ 30 ÷ 6
⑤ 65 ÷ 9	⑥ 28 ÷ 4	⑦ 8 ÷ 6	⑧ 9 ÷ 3
⑨ 32 ÷ 2	⑩ 96 ÷ 4	⑪ 78 ÷ 6	⑫ 98 ÷ 7
⑬ 75 ÷ 5	⑭ 60 ÷ 4	⑮ 76 ÷ 2	⑯ 84 ÷ 6
⑰ 85 ÷ 4	⑱ 66 ÷ 5	⑲ 94 ÷ 7	⑳ 73 ÷ 6
㉑ 95 ÷ 8	㉒ 68 ÷ 3	㉓ 61 ÷ 3	㉔ 83 ÷ 4

2 Let's calculate the following in vertical form. Let's confirm the answer.

① 75 ÷ 4	② 53 ÷ 3	③ 86 ÷ 3	④ 74 ÷ 6

3 We want to divide 55 sheets of colored paper so that 7 children receive equally. How many sheets will each child receive? How many sheets will remain?

4 We want to divide 63 candies so that each child receives 5 candies. How many children can receive the candies? How many candies will remain?

5 Let's calculate the following in vertical form.

① 314 ÷ 2	② 447 ÷ 3	③ 775 ÷ 5	④ 588 ÷ 4
⑤ 825 ÷ 3	⑥ 912 ÷ 8	⑦ 891 ÷ 3	⑧ 984 ÷ 4
⑨ 196 ÷ 7	⑩ 783 ÷ 9	⑪ 199 ÷ 4	⑫ 116 ÷ 2
⑬ 779 ÷ 8	⑭ 145 ÷ 3	⑮ 289 ÷ 5	⑯ 194 ÷ 3
⑰ 840 ÷ 6	⑱ 810 ÷ 3	⑲ 915 ÷ 7	⑳ 654 ÷ 5
㉑ 618 ÷ 2	㉒ 836 ÷ 4	㉓ 736 ÷ 7	㉔ 975 ÷ 9

6 We want to divide 417 sheets of colored paper so that 3 groups receive equally. How many sheets of paper will each group receive?

7 Karen is reading a book with 224 pages. She reads 8 pages a day. How many days will it take for her to finish reading the book?

Arrangement of Data

→ pp.79 ～ 85

1 The table on the right summarizes the injuries that occurred at school last month, focusing on the places where the injuries occurred and the kind of the injuries. Let's answer the followings about this table.

Places and kind of injuries (children)

Place \ Kind	Cut	Bruise	Scratch	Sprain	Total
Playground	1	4		1	
Corridor	2	3	2	0	
Classroom	4	1	3	0	
Gymnasium	1	2	4	2	
Stairs	0	1	2	0	
Total	8		16		

① What are the numbers of children for the following?

ⓐ The number of children with bruise at the corridor

ⓑ The number of children with scratch at the gymnasium

② Let's fill in the blanks in the table with numbers.

③ What kind of injury occurred most frequently?

④ Where did injuries occur most frequently?

⑤ How many children were injured in total at school last month?

2 Haruma investigated whether his classmates had a dog or a cat. The following information was gathered.

- There are 37 students in Haruma's class.
- 24 children have a dog.
- 11 children have a dog and a cat.
- 18 children do not have a cat.
- 8 children have a cat, but not a dog.

Let's answer the following.

Animals that children have (children)

<table>
<tr><td colspan="2" rowspan="2"></td><td colspan="2">Dog</td><td rowspan="2">Total</td></tr>
<tr><td>○</td><td>×</td></tr>
<tr><td rowspan="2">Cat</td><td>○</td><td></td><td></td><td></td></tr>
<tr><td>×</td><td></td><td></td><td></td></tr>
<tr><td colspan="2">Total</td><td></td><td></td><td></td></tr>
</table>

(○…have ×…do not have)

① Let's write down in the table the numbers of children based on what you have gathered. Let's fill in the blanks in the table with numbers too.

② How many children do not have either a dog or a cat?

③ How many children only have a dog? How many children do not have a dog?

④ How many children have a cat?

8 Division by 2-digit Numbers

→ pp.88 ~ 104

1 Let's calculate the following in vertical form.

① $84 \div 42$ ② $48 \div 12$ ③ $66 \div 22$ ④ $29 \div 13$
⑤ $67 \div 21$ ⑥ $96 \div 31$ ⑦ $52 \div 13$ ⑧ $84 \div 28$
⑨ $90 \div 37$ ⑩ $75 \div 15$ ⑪ $72 \div 14$ ⑫ $85 \div 17$

2 We want to cut a ribbon that is 65cm long into pieces with the length of 16cm. How many pieces with the length of 16cm can we get? How many cm will remain?

3 Let's calculate the following in vertical form.

① $210 \div 42$ ② $243 \div 81$ ③ $218 \div 38$
④ $343 \div 53$ ⑤ $539 \div 74$ ⑥ $282 \div 47$
⑦ $309 \div 31$ ⑧ $742 \div 77$ ⑨ $620 \div 69$

4 We want to divide 135 sheets of colored paper so that 18 children will receive equally. How many sheets of paper will each child receive? How many sheets will remain?

5 We want to divide 230 balls so that 1 box contains 40 balls. How many boxes will be filled? How many balls will remain?

6 Let's calculate the following in vertical form.

① $450 \div 18$ ② $551 \div 46$ ③ $855 \div 57$
④ $391 \div 23$ ⑤ $432 \div 27$ ⑥ $965 \div 36$
⑦ $924 \div 46$ ⑧ $648 \div 16$ ⑨ $963 \div 48$
⑩ $729 \div 24$ ⑪ $765 \div 76$ ⑫ $688 \div 34$

Perpendicular, Parallel, and Quadrilaterals

→ pp.112 ~ 137

1 Let's draw the following straight lines.

① a perpendicular straight line that goes through point A to straight line *a*

② a straight line that is parallel to line *b* that goes through point B

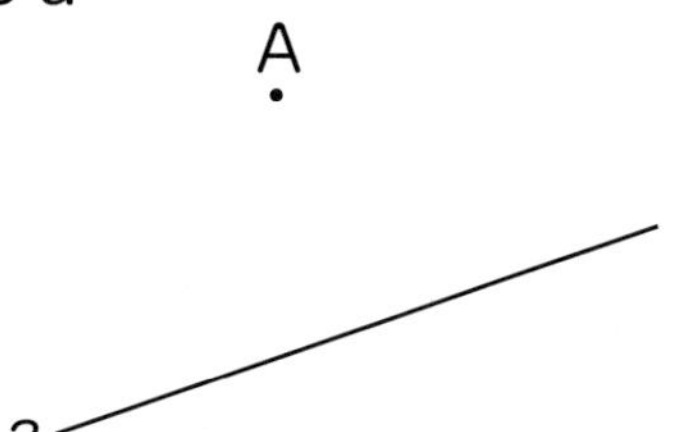

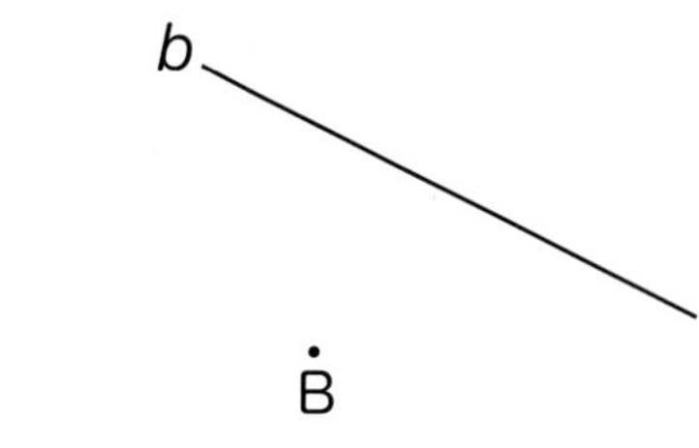

2 Let's answer the questions about the rhombus on the right.

① Which side is parallel to side AB?

② How many degrees is the size of angle D?

③ Let's draw the same rhombus as this one.

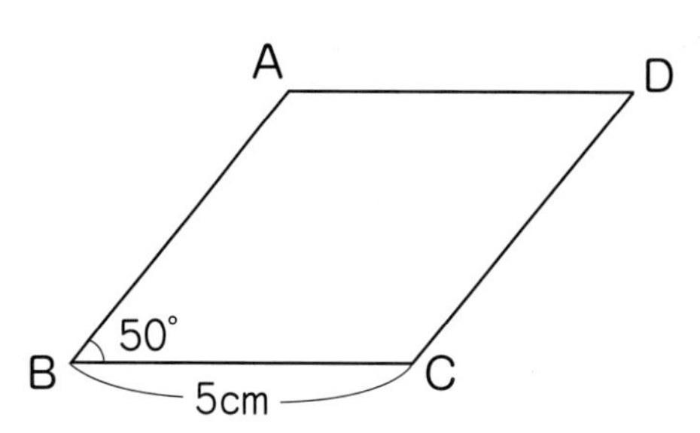

3 Let's choose all quadrilaterals that have the properties shown in ①~④ below.

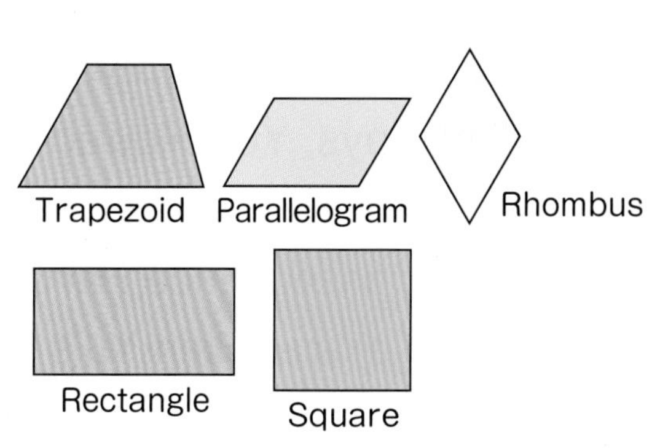

① a quadrilateral where two diagonals intersect perpendicularly

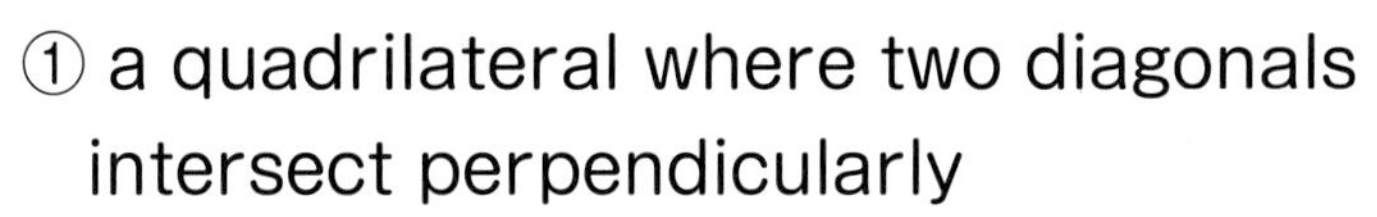

② a quadrilateral where two diagonals have the same length

③ a quadrilateral where two diagonals intersect perpendicularly and have the same length

④ a quadrilateral in which two diagonals are divided into 2 equal parts at the point where they intersect

4 What kind of quadrilateral is drawn by connecting the points A → B → C → D → A with straight lines in the diagram on the right?

①

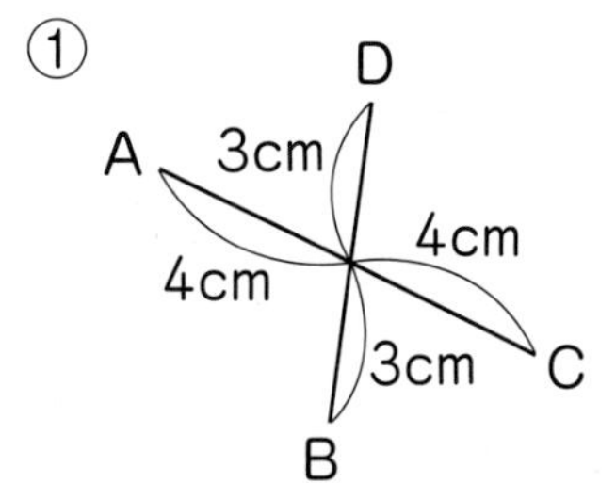

②

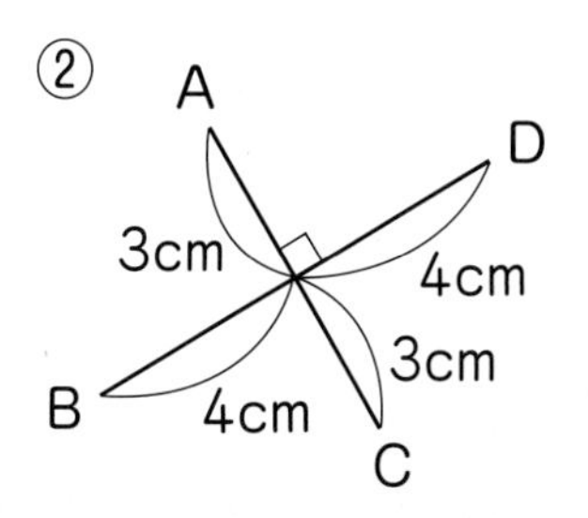

Since born, how many seconds?

Today is Haruto's birthday. He became 10 years old. Yu and Haruto are wondering how many seconds have passed since Haruto's birth to this date.

I turned 10 years old today! How many seconds are there since I was born?

Let's think in order.
First, 1 minute = 60 seconds.
As 1 hour = 60 minutes,
$60 \times 60 = 3600$, therefore,
1 hour has 3600 seconds.

① The figure on the right shows the process considering a year as 365 days, Yu followed to find out the number of seconds passed since Haruto's birth. Let's explain Yu's idea.

$60 \times 60 = 3600$

$3600 \times 24 = 86400$

$86400 \times 365 = 31536000$

$31536000 \times 10 = 315360000$

② Between which ages will there be more than 1 billion (10億) seconds passed since Haruto was born? Let's choose from Ⓐ～Ⓓ and explain the reason using math expressions and words.

Ⓐ Between 10 and 20 years old　Ⓑ Between 20 and 30 years old

Ⓒ Between 30 and 40 years old　Ⓓ Between 40 and 50 years old

The number representing the time since I was born in seconds became so large!

How many trees?

In Manganji (Niigata CIty, Niigata Prefecture), we can find Hasaki, which is a tree that is planted along the rice fields in order to dry rice after mowing.
Let's consider the case in which these trees are planted in a line like this.

Hasaki in Manganji
(Niigata City, Niigata Pref.)

① The trees are planted every 4m as shown below. When the length of the road is 80m, how many trees are there?

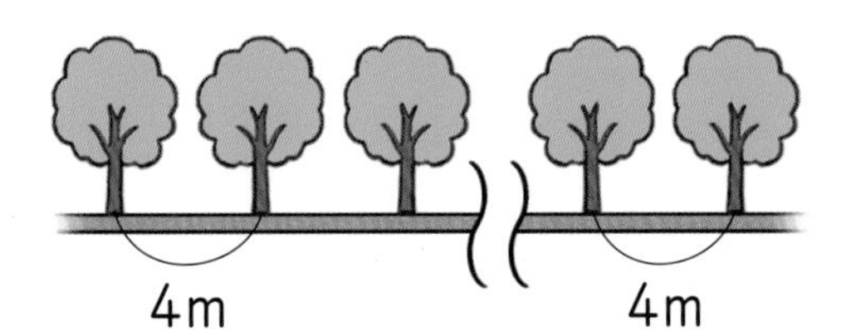

② When the trees are planted every 2m and the length of the road is 1000m, how many trees are there?

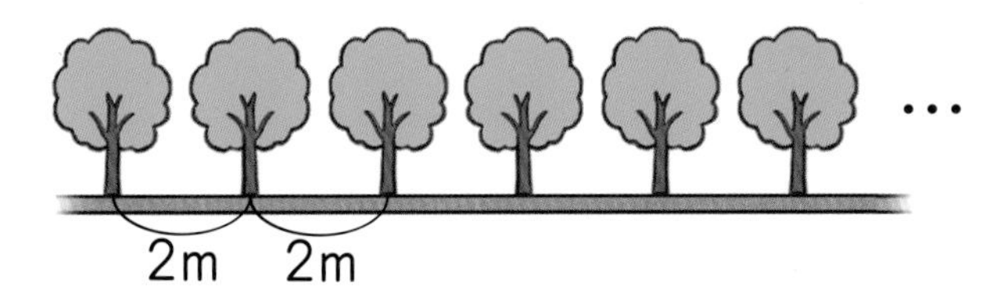

③ If trees are planted on both sides of the road in the situation of ②, what will the total number of trees be?

How many passengers?

Four children are talking about the number of passengers that were riding on the train. Let's think about the number of passengers.

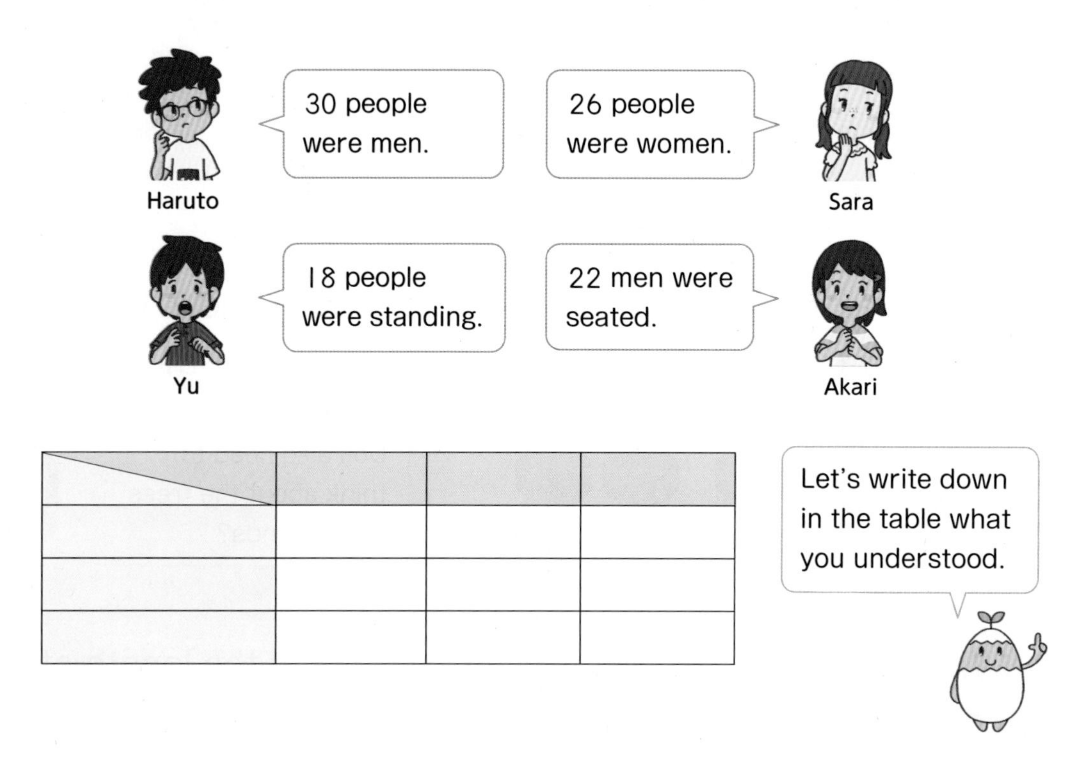

30 people were men.
Haruto
26 people were women.
Sara
18 people were standing.
Yu
22 men were seated.
Akari
Let's write down in the table what you understood.

① How can we find out the total number of passengers on this train? Explain your idea.

② How can we find out the number of men standing on this train? Explain your idea.

③ How can we find out the number of women standing on this train? Explain your idea.

What kind of quadrilateral can you make?

Let's think about the diagram on the right. What quadrilateral will be created if you connect the following four points with straight lines?
Let's think and draw on the diagram below. Let's discuss the reason for each process.

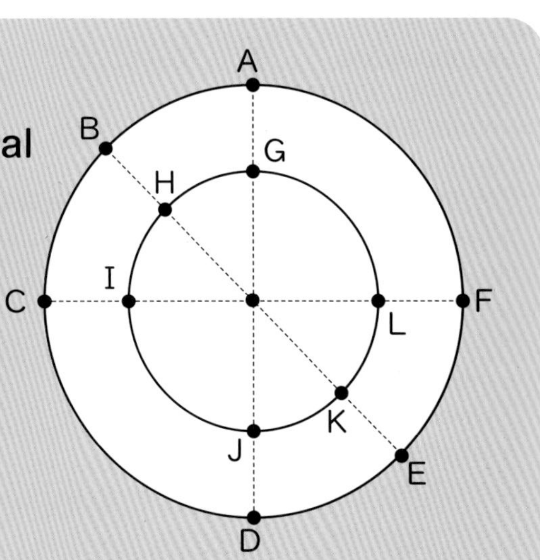

① point B, point C, point E, point F

② point G, point I, point J, point L

③ point G, point C, point J, point F

④ point A, point H, point D, point K

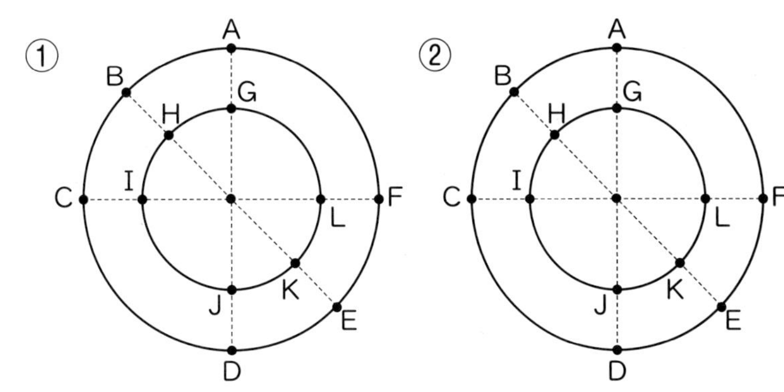

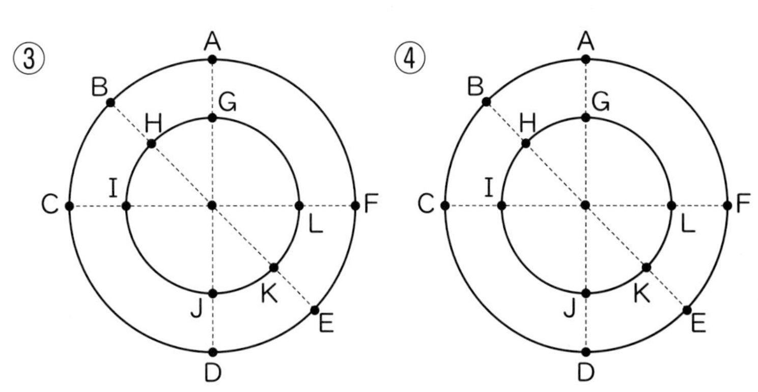

Answers

[Supplementary Problems]

1 Large Numbers → p.142

1 ① 30500000000 ② 47069000000
③ 321000000000000
④ 50098000000000

2 ① 38 ② 99 million ③ 29 ④ 100

3 ① 300 thousand ② 500 billion ③ 4 billion
④ 70 billion ⑤ 800 million ⑥ 500 billion

4 ① < ② < ③ > ④ >

5 The largest number…33210
The smallest number…10233

6 ① 6 million 410 thousand
② 6 billion 700 million ③ 450 trillion
④ 6 billion 800 million

2 Line Graphs → p.143

1 ① 9℃ ② 1 p.m. and 3 p.m.
③ 16℃, 2 p.m.
④ Between 10 a.m. and 11 a.m. ⑤ 3℃ ⑥ 8℃

2 Ⓑ, Ⓓ

3

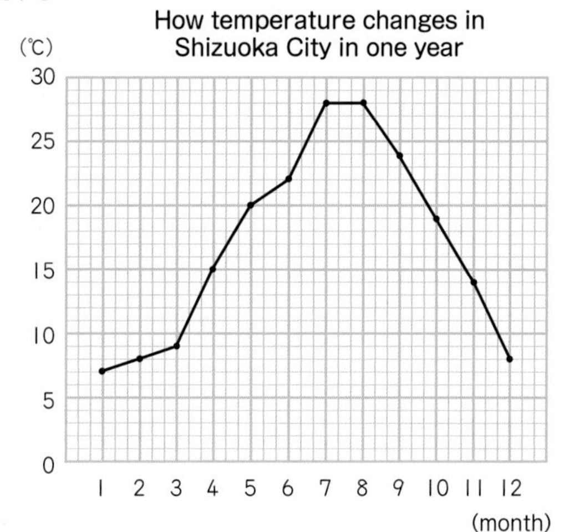

3 Division → p.144

1 ① ⓐ 4 ⓑ 4 ⓒ 9
② ⓐ 2 ⓑ 2 ⓒ 6
③ ⓐ 3 ⓑ 3 ⓒ 3
④ ⓐ 2 ⓑ 2 ⓒ 1
⑤ ⓐ 6 ⓑ 6 ⓒ 1
⑥ ⓐ 3 ⓑ 3 ⓒ 9

2 ① 30, 6 ② 7, 4 ③ 250, 5

3 ① 4, 2 ② 9, 3 ③ 16, 8

4 Angles → p.145

1 ① 90 ② 2 ③ 360

2 ① 70° ② 55° ③ 220° ④ 300°

3 ⓐ 105° ⓑ 75° ⓒ 105°

4 ① ② ③

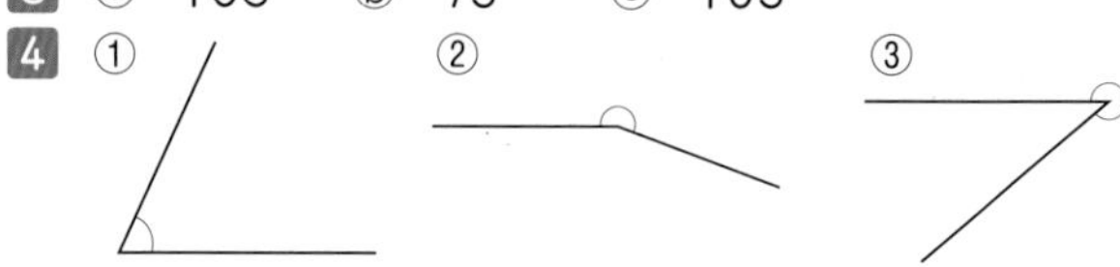

5 ① ⓐ 135° ⓑ 285° ② ⓒ 15°

6 Division by 1-digit Numbers → p.146

1 ① 8 remainder 6 ② 6 remainder 2
③ 7 remainder 1 ④ 5 ⑤ 7 remainder 2
⑥ 7 ⑦ 1 remainder 2 ⑧ 3 ⑨ 16
⑩ 24 ⑪ 13 ⑫ 14 ⑬ 15 ⑭ 15
⑮ 38 ⑯ 14 ⑰ 21 remainder 1
⑱ 13 remainder 1 ⑲ 13 remainder 3
⑳ 12 remainder 1 ㉑ 11remainder 7
㉒ 22 remainder 2 ㉓ 20 remainder 1
㉔ 20 remainder 3

2 ① 18 remainder 3, $4 \times 18 + 3 = 75$
② 17 remainder 2, $3 \times 17 + 2 = 53$
③ 28 remainder 2, $3 \times 28 + 2 = 86$
④ 12 remainder 2, $6 \times 12 + 2 = 74$

3 Each child receives 7 sheets and 6 sheets remain

4 12 children, 3 candies remain

5 ① 157 ② 149 ③ 155 ④ 147
⑤ 275 ⑥ 114 ⑦ 297 ⑧ 246
⑨ 28 ⑩ 87 ⑪ 49 remainder 3
⑫ 58 ⑬ 97 remainder 3 ⑭ 48 remainder 1
⑮ 57 remainder 4 ⑯ 64 remainder 2
⑰ 140 ⑱ 270 ⑲ 130 remainder 5
⑳ 130 remainder 4 ㉑ 309 ㉒ 209
㉓ 105 remainder 1 ㉔ 108 remainder 3

6 139 sheets

7 28 days

7 Arrangement of Data

→ p.147

1 ① ⓐ 3 children ⓑ 4 children

②

Places and kind of injuries (children)

Kind / Place	Cut	Bruise	Scratch	Sprain	Total
Playground	1	4	5	1	11
Corridor	2	3	2	0	7
Classroom	4	1	3	0	8
Gymmasium	1	2	4	2	9
Stairs	0	1	2	0	3
Total	8	11	16	3	38

③ Scratch ④ Playground ⑤ 38 children

2 ①

Animals that children have (children)

		Dog		Total
		○	×	
Cat	○	11	8	19
	×	13	5	18
Total		24	13	37

② 5 children ③ 13 children, 13 children
④ 19 children

8 Division by 2-digit Numbers

→ p.148

1 ① 2 ② 4 ③ 3 ④ 2 remainder 3
⑤ 3 remainder 4 ⑥ 3 remainder 3 ⑦ 4 ⑧ 3
⑨ 2 remainder 16 ⑩ 5 ⑪ 5 remainder 2 ⑫ 5

2 4 pieces, 1cm remains

3 ① 5 ② 3 ③ 5 remainder 28 ④ 6 remainder 25
⑤ 7 remainder 21 ⑥ 6 ⑦ 9 remainder 30
⑧ 9 remainder 49 ⑨ 8 remainder 68

4 7 sheets, 9 sheets remain

5 5 boxes, 30 balls remain

6 ① 25 ② 11 remainder 45 ③ 15 ④ 17
⑤ 16 ⑥ 26 remainder 29 ⑦ 20 remainder 4
⑧ 40 remainder 8 ⑨ 20 remainder 3
⑩ 30 remainder 9 ⑪ 10 remainder 5
⑫ 20 remainder 8

9 Perpendicular, Parallel, and Quadrilaterals

→ p.149

1 ①

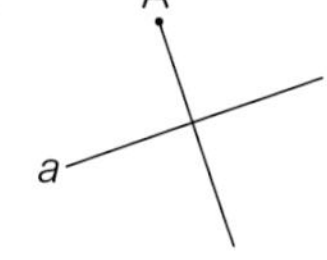

②

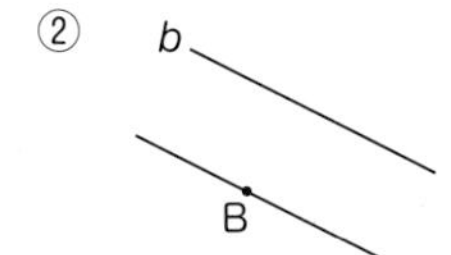

2 ① side DC ② 50° ③ (omitted)

3 ① Rhombus, Square ② Rectangle, Square
③ Square
④ Parallelogram, Rhombus, Rectangle, Square

4 ① Parallelogram ② Rhombus

[Let's deepen.]

Since born, how many seconds?

→ p.150

① (omitted)
② Ⓒ (explanation omitted)

How many trees?

→ p.151

① 21 trees ② 501 trees
③ 1002 trees

How many passengers?

→ p.152

	Man	Woman	Total
Standing	8	10	18
Sitting	22	16	38
Total	30	26	56

① Add the number of women and the number of men.
② Subtract the number of seated men from the total number of men.
③ Subtract ② from the total of people standing.

What kind of quadrilateral can you make?

→ p.153

① Rectangle ② Square
③ Rhombus ④ Parallelogram

which we learned in this textbook

Various quadrilaterals

→ To be used in pages 112 and 113.
Please cut these out for use.

Angles

→ To be used in pages 48 and 49.
Please cut these out for use.

Memo

Memo

Memo

Editors of Original Japanese Edition

[Head of Editors]

Shin Hitotsumatsu (Kyoto University), Yoshio Okada (Hiroshima University)

[Supervising Editors]

Toshiyuki Akai (Hyogo University), Toshikazu Ikeda (Yokohama National University), Shunji Kurosawa (The former Rikkyo University), Hiroshi Tanaka (The former Tsukuba Univ. Elementary School), Kosho Masaki (The former Kokugakuin Tochigi Junior College), Yasushi Yanase (Tamagawa University)

[Editors]

Shoji Aoyama (Tsukuba Univ. Elementary School), Kengo Ishihama (Showa Gakuin Elementary School), Hiroshi Imazaki (Hiroshima Bunkyo University), Atsumi Ueda (Hiroshima University), Tetsuro Uemura (Kagoshima University), Yoshihiro Echigo (Tokyo Gakugei Univ. Setagaya Elementary School), Hisao Oikawa (Yamato University), Hironori Osawa (Yamagata University Graduate School), Tomoyoshi Owada (Shizuoka University), Nobuhiro Ozaki (Seikei Elementary School), Masahiko Ozaki (Kansai Univ. Elementary School), Kentaro Ono (Musashino University), Hiroshi Kazama (Fukui University), Michihiro Kawasaki (Oita University), Miho Kawasaki (Shizuoka University), Yoshiko Kambe (Tokai University), Yukio Kinoshita (Kwansei Gakuin Elementary School), Tomoko Kimura (Tamon Elementary School, Setagaya City), Satoshi Kusaka (Naruto University of Education), Kensuke Kubota (Naruo Higashi Elementary School, Nishinomiya City), Itsushi Kuramitsu (The former University of the Ryukyus), Maiko Kochi (Kounan Elementary School, Toshima City), Chihiro Kozuki (Daiyon. Elementary School, Hino City), Goto Manabu (Hakuoh University), Michihiro Goto (Tokyo Gakugei Univ. Oizumi Elementary School), Hidenori Kobayashi (Hiroshima Univ. Shinonome Elementary School), Akira Saito (Shibata Gakuen University Graduate School), Masahiko Sakamoto (The former Tokoha University Graduate School), Junichi Sato (Kunitachigakuen Elementary School), Hisatsugu Shimizu (Keio Yochisha Elementary School), Ryo Shoda (Seikei University), Masaaki Sugihara (University of the Sacred Heart, Tokyo), Jun Suzuki (Gakushuin Primary School), Shigeki Takazawa (Shiga University), Chitoshi Takeo (Nanzan Primary School), Hidemi Tanaka (Tsukuba Univ. Elementary School), Toshiyuki Nakata (Tsukuba Univ. Elementary School), Hirokazu Nagashima (Daisan. Elementary School, Kokubunji City), Minako Nagata (Futaba Primary School), Kiyoto Nagama (Hiyagon Elementary School, Okinawa City), Satoshi Natsusaka (Tsukuba Univ. Elementary School), Izumi Nishitani (Gunma University), Kazuhiko Nunokawa (Joetsu University of Education), Shunichi Nomura (Waseda University), Mantaro Higuchi (Kori Nevers Gakuin Elementary School), Satoshi Hirakawa (Showa Gakuin Elementary School), Kenta Maeda (Keio Yokohama Elementary School), Hiroyuki Masukawa (University of the Sacred Heart, Tokyo), Shoichiro Machida (Saitama University), Keiko Matsui (Hasuike Elementary School, Harima Town), Katsunori Matsuoka (Naragakuen University), Yasunari Matsuoka (Matsushima Elementary School, Naha City), Yoichi Matsuzawa (Joetsu University of Education), Satoshi Matsumura (Fuji Women's University), Takatoshi Matsumura (Tokoha University), Kentaro Maruyama (Yokohama National Univ. Kamakura Elementary School), Kazuhiko Miyagi (Homei Elementary School Affiliated with J.W.U), Aki Murata (University of California, Berkeley), Takafumi Morimoto (Tsukuba Univ. Elementary School), Yoshihiko Moriya (The former Kunitachigakuen Elementary School), Junichi Yamamoto (Oi Elementary School, Oi Town), Yoshikazu Yamamoto (Showa Gakuin Elementary School), Shinya Wada (Kagoshima University), Keiko Watanabe (Shiga University)

[Reviser]

Katsuto Enomoto (The former Harara Elementary School, Kagoshima City)

[Reviser of Special Needs Education and Universal Design]

Yoshihiro Tanaka (Teikyo Heisei University)

[Cover]

Photo : Tatsuya Tanaka (MINIATURE LIFE)

Design : Ai Aso (ADDIX)

[Text]

Design : Ai Aso, Takayuki Ikebe, Hanako Morisako, Miho Kikuma, Sawako Takahashi, Katsuya Imamura, Satoko Okutsu, Kazumi Sakaguchi, Risa Sakemoto (ADDIX), Ayaka Ikebe

[Illustrations]

Lico, Kozue Gomita, DOKOCHALUCHO, Ryoko Totsuka, Kinue Naganawa, Mayumi Nojima, B-rise

[Photo・Observation data]

Ritmo, Aflo, Pixta, Yamaguchi Tourism & Convention Association, Iemasa Yamahata(kogin.net), Akiha Ward Office,Niigata City